Entering Mentoring

Entering Mentoring

Christine Pfund
University of Wisconsin–Madison, Wisconsin Center for Education Research

Janet Branchaw
University of Wisconsin–Madison, Institute for Biology Education

Jo Handelsman
Yale University, Department of Molecular, Cellular, and Developmental Biology and Center for Scientific Teaching

Part of the
W.H. Freeman Entering Mentoring Book Series

www.whfreeman.com

Acquisitions Editor:	Beth Cole
Editor:	Sara Ruth Blake
Executive Marketing Manager:	John Britch
Cover and Text Designer:	Mark Ong, Side By Side Studios
Director of Production:	Valerie Zaborski
Composition:	Susan Riley, Side By Side Studios
Printing and Binding:	RR Donnelley

Entering Mentoring Series Editor: Christine Pfund, Wisconsin Center for Education Research, University of Wisconsin–Madison

Entering Mentoring Series Editor: Jo Handelsman, Department of Molecular, Cellular, and Developmental Biology and Center for Scientific Teaching, Yale University

Cover Image: Godong/Robert Harding World Imagery/Getty Images

Curricular materials from this book and other mentoring resources are available online at https://mentoringresources.ictr.wisc.edu and www.researchmentortraining.org

Library of Congress Control Number: 2014933772

ISBN-13: 978-1-4641-8490-1
ISBN-10: 1-4641-8490-9

W.H. Freeman & Company
41 Madison Avenue
New York, NY 10010
Houndmills, Basingstoke
RG21 6XS, England
www.whfreeman.com

Contents

3 Promoting Professional Development   45

4 Maintaining Effective Communication   57

5 Addressing Equity and Inclusion   63

6 Assessing Understanding   83

Foreword

Professor Jo Handelsman, Dr. Christine Pfund, and colleagues in the Wisconsin Program for Scientific Teaching at the University of Wisconsin had a very good idea in 2003: teach mentoring to graduate students and postdoctoral fellows. This book is the revised edition of an extremely useful and noteworthy aid to implement that idea, developed from research and experience, and presented here as a practical and extremely straightforward guide to address this important need.

Undergraduate research has become part of many students' undergraduate experience, but the mentoring of these students, although critically important, is often inadequate. Especially in research universities, daily laboratory contact is not with the faculty member in charge, but rather with graduate students and postdoctorals; students are the immediate mentors of these undergraduates. Unfortunately, the reality is that these student mentors often fly by the seat of their pants, lacking any training on the complex and diverse issues involved in mentoring. So to improve undergraduate research experiences, the authors recognized that something needed to be done to help mentors.

Ironically, training graduate students and postdoctorals to be effective mentors has an impact beyond improving today's interactions in the lab; it will improve the professorate of the future. Traditionally, effective mentoring is something not covered in graduate curricula; most faculty learn by on-the-job experience at best, and some never learn. Although interactions with students can be extremely complex and formative for the mentee, understanding the issues involved in being an effective mentor has rarely been part of training to become a scientist.

Handelsman and her colleagues recognized that accomplishing the present need for improved mentoring of student researchers and the longer-range impact of enhancing the training of the professorate of the future could be addressed by a straightforward set of activities: mentor training seminars and workshops. They also understood that these activities would be an add-on to normal curricula, and incorporated materials to assist the facilitators. This book and the others in this series are designed to make it easy to present these training experiences.

The material in this volume was created following interviews and focus groups, refined by practice at UW–Madison and then a group of research universities, and strongly driven by the authors' research on effective learning. This edition presents separate chapters focusing on seven key topics. Within each chapter are useful case studies and other handy information on the topic at hand. But beyond the basic information, the book provides extremely useful aids to effective presentation of the material in a variety of settings, and organizes the chapters around well-articulated learning objectives. Finally, the series is linked to a web resource, so information can be updated and supplemented with a growing collection of case studies and other appropriate materials.

In short, this book serves as an important gateway to a curriculum aimed at improving an extremely significant yet remarkably overlooked educational practice: effective mentoring. The book practices what it preaches; it provides useful active learning elements and activities in a form that is easy to apply to a variety of settings. It should be a go-to owner's manual for any educational institution in which advanced students mentor beginning students. It works well to help faculty become better mentors too.

Peter Bruns, PhD
Former Vice President for Grants and Special Programs at the Howard Hughes Medical Institute
Professor Emeritus of Genetics, Cornell University

Preface

Mentoring principles, not practices, are universal

Effective mentoring can be learned, but not taught. Most faculty learn to mentor by experimenting and analyzing success and failure, and many say that the process of developing an effective method of mentoring takes years, which is a reflection of the unique qualities, needs, and challenges presented by each mentee. A skilled mentor is guided by a reflective philosophy that directs examination of the mentee's changing needs and how best to address them, creating fluidity in the relationship. No book can prescribe a single "right" approach, but systematic analysis and discussion of mentoring generates a method for tackling the knotty challenges inherent in the job.

The goal of the curriculum outlined in this book is to accelerate the process of becoming an effective research mentor. The approach described provides mentors with an intellectual framework, an opportunity to experiment with various methods, and a forum in which to solve mentoring dilemmas with the help of their peers. The mentor-training process expands each mentor's knowledge through secondhand exposure to the experiences of the entire group, enabling participants to engage with as many mentoring experiences as each of them would typically handle in a decade. This process in turn enhances their readiness to work with diverse mentees and anticipate new situations. At the completion of the training, mentors will have articulated their own approach to mentoring and have a toolbox of strategies to draw upon when confronted with mentoring challenges.

Although no one can provide formulas, practices, or behaviors that will work in every mentoring situation, certain principles guide good mentoring. The principles that shape this curriculum are founded on research that has revealed how people learn and has identified the essential elements of environments shown to be most conducive to learning, productivity, and creativity.

Mentoring diversity, not sameness, is essential

An individual's performance in any endeavor is the product of a complex interaction involving innate ability, experience, confidence, education, and the nature of the performance environment.

Professional mentors can directly influence their mentees' performance by creating an environment that is conducive to achieving excellence and that fosters confidence, even in stressful situations. Setbacks are a source of stress that everyone experiences, and the mentee's response can be modulated by a mentor's intervention. A mentor's goal is to promote a mentee's growth and achievement. People build resilience and self-reliance through positive reinforcement coupled with the expectation of excellence. The most important message a mentor can send is faith in the mentee, a willingness to embrace diversity, and an eagerness to continually improve as a mentor. A theme implicit in this book's curriculum is that mentors may facilitate growth best when they work collaboratively with their mentees to continually reexamine and adjust to their individual needs. This process, followed by the mentee producing high-quality research, generates self-sustaining confidence for both.

Another aspect of creating an environment that is conducive to learning is being open to other ways of doing research and seeing the world, including the world of academia. The next generation of researchers will be more diverse than the last. Working with people who are different from ourselves can at times be frustrating and baffling, though also enlightening and deeply rewarding as we learn from one another. When given the opportunity to work with mentees from different backgrounds and with distinct perspectives, who may not share the characteristics we value most in ourselves, we may struggle to imagine them fitting the researcher mold. We are often surprised by the success of those who don't immediately fit in, and find that they may be the very people who bring a key new perspective or insight. Being a good mentor requires accommodating styles that differ from our own, thereby enhancing the diversity and the vibrancy of the scientific community.

Christine Pfund
Series Editor
University of Wisconsin–Madison

Jo Handelsman
Series Editor
Yale University

Acknowledgments

The Research Mentor Training Seminar, *Entering Mentoring*, was originally developed by the Wisconsin Program for Scientific Teaching with support from the Howard Hughes Medical Institute Professors Program (PI: Jo Handelsman; Handelsman, J., Pfund, C., Miller Lauffer, S., and Pribbenow, C. M. [2005], *Entering Mentoring: A Seminar to Train a New Generation of Scientists,* Madison, WI: University of Wisconsin Press). The work was adapted for use across the natural and behavioral sciences, engineering, and mathematics disciplines with funding from the National Science Foundation (Grant # 0717731; PI: Christine Pfund) and implemented through the Center for the Integration of Research, Teaching, and Learning (CIRTL), its Delta Program in Research, Teaching, and Learning, and the Institute for Biology Education at the University of Wisconsin–Madison.

We would like to thank those from the University of Wisconsin–Madison who contributed to the development of the materials included in this multidisciplinary curriculum: Robert Beattie, Kimberly D'Anna, Amy Fruchtman, Andrew Greenberg, David Griffeath, Eric Hooper, Erin Jonaitis, Robert Mathieu, David McCullough, Trina McMahon, Brad Postle, Raelyn Rediske, Manuela Romero, Ashley Shade, David Wassarman, and Tehshik Yoon.

We thank Sarah Miller for her contributions to the development of the original *Entering Mentoring* curriculum and Christine Pribbenow for her leadership in evaluation during the development, implementation, and revision of this curriculum.

We thank those who dedicated time and effort to developing and refining the current structure of this new version of *Entering Mentoring*, including Pam Asquith, Gail Coover, Stephanie House, and Kim Spencer.

Curriculum Overview

The resources in this section will help you implement research mentor training in your own educational setting.

Content

The content of each session in this curriculum is designed to address the key concerns and challenges identified by research mentors. The sample syllabus shown here presents the topics in the recommended order. However, the sessions can easily be mixed and matched to create customized workshops.

Sample Entering Mentoring Syllabus

Sessions	Topics	Assignments Due	Readings
Week 1	Introduction to Mentor Training		Handelsman, "Mentoring: Learned, Not Taught"
Week 2	Aligning Expectations	Draft mentoring strategy or philosophy Description of mentee's research project	
Week 3	Promoting Professional Development	Draft mentoring compact/contract	
Week 4	Maintaining Effective Communication	Review of Individual Development Plans (IDPs)	
Week 5	Addressing Equity and Inclusion	Reflection on differences and how they affect the research experience	Fine and Handelsman, "Benefits and Challenges of Diversity"
Week 6	Assessing Understanding	Revised compacts and IDPs	
Week 7	Fostering Independence	Summary of discussion with your own mentor	Hughes, "Mentoring Research Writers"
Week 8	Cultivating Ethical Behavior	Look over the general ethics guidelines for your discipline. Be prepared to talk about how they apply to you and your work. Bring a copy of them to class.	
Week 9	Articulating Your Mentoring Philosophy and Plan	Revised mentoring philosophy	

Each of these topics is critical for mentoring, yet these divisions are, at some level, artificial and overlapping. However, focusing each session on one topic allows mentors to delve more deeply into that topic. Facilitators are encouraged to read through all the materials ahead of time so they can highlight linkages between topics throughout the training. In addition to general content about research mentoring, all of the case studies and some of the discussion questions draw specific attention to the unique circumstances and challenges in mentoring trainees working in science, technology, engineering, mathematics, and medical (STEMM) research. Additional materials for these topic areas in various disciplines are available at http://researchmentortraining.org and https://mentoring resources.ICTR.wisc.edu.

This curriculum is an updated version of *Entering Mentoring: A Seminar to Train a New Generation of Scientists*, created by Jo Handelsman, Christine Pfund, Sarah Miller Lauffer, and Christine Pribbenow. A PDF version of the original book, which focuses on training mentors of undergraduate researchers in the biological sciences, is available at http://www.hhmi.org/sites/default/files/Educational%20Materials/Lab%20Management/entering_mentoring.pdf. The content of the multidisciplinary curriculum presented here is based on the original curriculum and on materials that were subsequently created for training mentors across STEMM. It incorporates the breadth of these training materials and therefore serves as the comprehensive core framework for the entire Entering Mentoring Series.

Audience

This curriculum is designed for those who wish to implement mentorship development programs for academic research mentors in STEMM. It is tailored for the primary mentors of undergraduate researchers in any of the STEMM disciplines. *Entering Research*, a parallel training for undergraduate research mentees, is also available through W.H. Freeman.

Mentor-Mentee Relationship Parameters

STEMM research settings are often communal, with multiple researchers at various stages of their careers sharing tight lab space and working on overlapping projects. Regular research team meetings are standard and provide a setting in which feedback and mentoring can be provided. Mentors and mentees often have financial connections to each other, either through a dependence on shared equipment and space or through a direct employer-employee relationship in which researchers are funded as part of a program. In many cases, undergraduate researchers receive course credit for their research.

While the individual activities included in the curriculum may focus on a specific type of research or a specific aspect of the mentoring relationship, the curriculum as a whole is designed to include activities relevant to a broad range of mentors across diverse areas of research.

Format

The format of this research mentor-training program is based on the experience of faculty and staff who implemented the *Entering Mentoring* curriculum at UW–Madison. These facilitators have learned that the best results come from keeping an open discussion format to allow for participants' diverse experiences to be integrated into the training. Simply asking the mentors a few guiding questions typically leads to vigorous discussion. The case studies and reading materials can provide a tangible starting point, and the mentors often move quickly from these hypothetical examples to their own experiences with trainees. In fact, facilitators are encouraged to use the mentoring challenges expressed by participants in place of the provided case studies. The training is most effective with mentors who are currently working with research trainees because participants can immediately implement ideas derived from the training sessions. In particular, starting each session with reflections about changes participants have made in their mentoring practices since the previous session has proven very effective.

The preceding sample syllabus provides an example of how the training might be structured as nine 1-hour sessions. Ideally, it is best to hold the first session with research mentors *before* they begin working with their research mentees. While the spacing between these sessions is flexible, former participants have found that separating them by 1 week allows adequate time for reflection and implementation.

Implementation: Facilitating Research Mentor Training

Facilitating research mentor training is not the same as teaching it. The role of the facilitator is to enable participants to take ownership of their own learning by helping them engage in self-reflection, shared discovery, and learning. The facilitator works to build a community of mentors learning together, with the common goal of becoming more effective in their mentoring relationships. Importantly, the role of the facilitator is to help others work through their thoughts and ideas; it is not to be the expert on mentoring. Facilitators walk a fine line between facilitator and participant. Group members will look to the facilitator for guidance and structure, and often the facilitator's experiences and ideas can enhance the discussion; however, the facilitator's own challenges and solutions in mentoring should not dominate and become the focus of the discussion.

Being an effective facilitator is the key to helping the research mentors meet the learning objectives and become more successful mentors. To this end, we include a brief facilitator guide in the Appendix that contains additional information, tips, and tools for facilitation.

Using This Guidebook to Facilitate Weekly Sessions

This guidebook contains facilitator instructions and materials for each of the nine sessions outlined in the sample syllabus. Each session is organized as follows:

1. Introduction
2. Learning objectives
3. "Overview of Activities" table
4. Facilitation guide, including recommended session length, materials needed, objectives in detail, and post-session assignments
5. Activities, case studies, handouts, readings, and mentoring tools

Facilitators should prepare for each session by copying the learning objectives, case studies, worksheets, mentoring tools, and readings for each mentor in the group. Alternatively, all the materials can be copied at the start of the sessions and distributed at the first meeting or posted on a website. The specific themes and objectives for each session are included at the beginning of the materials. Facilitators might consider asking participants to review the themes and learning objectives at the beginning of each session, or to review the objectives and themes after a few weeks to check their progress.

Guiding discussion questions and notes for group facilitators are also included in each session plan. Time estimates for activities and facilitated discussions for each session are indicated in parentheses and can be adjusted at the facilitator's discretion. The facilitator notes provide directive signposts to support the facilitation process as described below:

ACTIVITY   Participants are to engage in some process on their own, in small groups, or as a large group.

TELL   Information that follows needs to be shared with the whole group.

ASK   A specific question needs to be put to the group.

NOTE   Some particular issue or content needs to be emphasized.

DISCUSS   A broader discussion, usually supported by guiding questions, needs to occur. Sometimes more discussion questions are provided than can reasonably be addressed in the time allotted for the activity or group discussion, but the questions suggested for the case studies in this training are based on the experiences of past facilitators.

Additional questions for the case studies, as well as additional case studies, are available at the "Build Your Own" section of the "Curriculum Options" tab at www.researchmentortraining.org.

A set of mentoring tools is provided at the end of several sessions. These tools are from the *Entering Research* curriculum for use by mentors with their mentees.

Grading

This curriculum emphasizes experiential learning and the integration of knowledge—drawn from reflection, discussion, readings, and activities—with practice. The curriculum can be implemented as a seminar/course for credit. Under this structure, attending the sessions, doing the assignments, and participating in the activities results in an A for the class. This approach allows each participant to invest personally in the learning experience and develop a unique and authentic identity as a mentor. In some cases, the course has been offered as a practicum in which credit is given for participating in the training while simultaneously mentoring an undergraduate researcher.

Assessment

Upon completion of the training, facilitators might consider asking participants to complete an evaluation survey to collect feedback on the training sessions themselves, on the skills of the facilitator, and on the knowledge, skill, and attitude gains of participants. Some example assessment instruments can be found at the weblinks notes referenced previously.

Learning Objectives

Below are measurable learning objectives for each mentor training topic. At the beginning of each chapter, a table is presented to align the training activities with the learning objectives.

Introduction

Mentors will have the knowledge and skills to

1. Learn about other mentors in the group to begin building a learning community
2. Reflect on group dynamics and ways to make the group functional
3. Establish ground rules for participation
4. Identify qualities of good research projects for undergraduate mentees
5. Prepare to establish effective research mentoring relationships with mentees

Aligning Expectations

Mentors will have the knowledge and skills to

1. Design and communicate clear goals for the research project
2. Listen to and consider the expectations of their mentee in the mentoring relationship
3. Consider how personal and professional differences may impact expectations
4. Clearly communicate expectations for the mentoring relationship
5. Align mentee and mentor expectations

Promoting Professional Development

Mentors will have the knowledge and skills to

1. Identify the roles mentors play in the overall professional development of their mentees
2. Develop a strategy for guiding professional development using a written format
3. Initiate and sustain periodic conversations with mentees on professional goals and career development objectives and strategies
4. Engage in open dialogue on balancing the competing demands, needs, and interests of mentors and mentees (e.g., research productivity, grant funding, creativity, independence, career preference decisions, nonresearch activities, personal development, and work-family balance)

Maintaining Effective Communication

Mentors will have the knowledge and skills to

1. Provide constructive feedback
2. Use multiple strategies for improving communication (in person, at a distance, across multiple mentees, and within appropriate personal boundaries)
3. Engage in active listening
4. Communicate effectively across diverse dimensions, including varied backgrounds, disciplines, ethnicities, and positions of power

Addressing Equity and Inclusion

Mentors will have the knowledge and skills to

1. Increase understanding of equity and inclusion and their influence on mentor-mentee interactions
2. Recognize the impact of conscious and unconscious assumptions, preconceptions, biases, and prejudices on the mentor-mentee relationship and acquire skills to manage them
3. Identify concrete strategies for learning about and addressing issues of equity and inclusion

Assessing Understanding

Mentors will have the knowledge and skills to

1. Assess their mentee's understanding of core concepts and processes and ability to develop and conduct a research project, analyze data, and present results
2. Identify reasons for a lack of understanding, including expert-novice differences
3. Use multiple strategies to enhance mentee understanding across diverse disciplinary perspectives

Fostering Independence

Mentors will have the knowledge and skills to

1. Define independence, its core elements, and how those elements change over the course of a mentoring relationship
2. Employ various strategies to build their mentee's confidence, establish trust, and foster independence
3. Create an environment in which mentees can achieve goals

Cultivating Ethical Behavior

Mentors will have the knowledge and skills to

1. Articulate ethical issues they need to discuss with their mentees
2. Clarify their roles as teachers and role models in educating mentees about ethics
3. Manage the power dynamic inherent in the mentoring relationship

Articulating Your Mentoring Philosophy and Plan

Mentors will have the knowledge and skills to

1. Reflect on the mentor-training experience
2. Reflect on intended behavioral or philosophical changes
3. Articulate an approach for working with mentees in the future

Introduction to Mentor Training

Introduction

Establishing group dynamics and laying the ground rules are perhaps two of the most important steps to launching a successful mentor-training program. Once established, these parameters help ensure mentors engage in shared learning of ways to become more effective mentors.

Learning Objectives

Mentors will

1. **Learn about other mentors in the group to begin building a learning community**
2. **Reflect on group dynamics and ways to make the group functional**
3. **Establish ground rules for participation**
4. **Identify qualities of good research projects for undergraduate mentees**
5. **Prepare to establish effective research mentoring relationships with mentees**

Overview of Activities for the "Introduction to Mentor Training" Session

	Learning Objectives	Core Activities
1	Learn about other mentors in the group to begin building a learning community	Preintroductory online social networking (Activity #1) Introductory activity (Activity #2)
2	Reflect on group dynamics and ways to make the group functional	Constructive and destructive group behaviors (Activity #3)
3	Establish ground rules for participation	Give or generate group ground rules (Activity #4)
4	Identify qualities of good research projects for undergraduate mentees	Brainstorm elements of effective research projects (Activity #5)
5	Prepare to establish effective research mentoring relationships with mentees	Share ideas for introductory activities in which to engage with mentees (Activity #6) Write mentee project description and draft mentoring philosophy (Postsession Assignment)

FACILITATION GUIDE

Recommended Session for Introduction to Mentor Training (60 minutes)

Materials Needed for the Session

- Table tents (can use folded index cards) and markers
- List of participants
- Chalkboard, whiteboard, or flip chart
- Handouts:
  - Copies of introduction and learning objectives for Introduction to Mentor Training (page 9)
  - Copies of "Constructive and Deconstructive Group Behaviors" (page 15)
  - Copies of "Questions to Consider When Writing a Mentoring Philosophy" (page 16)
  - Copies of the reading, "Mentoring: Learned, Not Taught" (pages 17–25)

Objective 1: Learn about other mentors in the group to begin building a learning community (15 min)

ACTIVITY #1: Preintroductory Activity

ASK ▶ Before the first training session, you might consider asking mentors to join a private online social networking community. This will enable them to become acquainted before the training, may enhance rapport during the sessions, and can provide an opportunity to build connections during and after the training. This is especially encouraged if mentors are from different institutions. Instruct mentors to create their own profiles and share basic information about themselves (name, title, department, area of research interest, mentoring experience). Alternatively, basic information or biographies could be collected from each participating mentor and distributed to the group before the training via email.

ACTIVITY #2: Introductory Activity (15 min)

ASK ▶ Invite participants to engage in an introductory activity from pages 13–14, or use one from your own experience.

Objective 2: Reflect on group dynamics and ways to make the group functional (10 min)

ACTIVITY #3: Building Constructive Group Dynamics (10 min)

ASK ▶ Ask participants to choose what is their own most constructive and destructive group behavior from the list of constructive and destructive group behaviors on page 15. Ask participants to write them on the back of their table tent or name tag.

TELL ▶ Tell each participant to explain their choices to the larger group.

DISCUSS ▶ Engage participants in a conversation about ways to handle destructive group behavior. For example, ask participants what facilitators and other participants should do if someone starts to dominate the conversation or completely withdraws from the discussion.

OR

ASK ▶ Ask participants to brainstorm a list of good and bad group behaviors and brainstorm ways to address these behaviors if they arise in the group.

NOTE ▶ These exercises help provide the group with a vocabulary so they can name these behaviors when they arise in themselves and others. It provides a lighthearted and nonthreatening way for participants to help each other stay on track and a nice segue to discussing communication.

Objective 3: Establish ground rules for participation (10 min)

ACTIVITY #4: Introduction to Research Mentor Training and Logistics

TELL ▶ (5 min) Provide participants with details about process and logistics for the program.
- Talk about the important roles research mentors play and the need for research mentor training.
- Explain how readings and assignments will be distributed and collected for the course.
  - If there is a course website, provide information about how to access and use the website.
  - You may ask participants to submit assignments ahead of time and compile them for everyone to read or simply bring copies of their assignments to class to share.
- Confidentiality is important to the process, and everything discussed in person or shared via website or email will remain confidential. Participants will also need to vote on whether they want their names removed from compiled assignments when work is shared with the entire class.
- Address how to enroll in the seminar/course if participants are taking it for credit. Typically, work is graded as a "contract A," which means if you show up, do the assignments, and participate, you will get an A.
- Activities and assignments are designed for mentors actively working with mentees. Research mentors who participate and are not mentoring concurrently can use their past experience in the discussions and adapt the assignments to plan for future mentoring relationships.

FACILITATION GUIDE

DISCUSS ▶ (5 min) Have the participants discuss expectations for attendance and reach consensus about how to deal with missed classes.

- How does the group want to handle participants missing class?
- Do we want participants to email the whole group or just the facilitator if they are going to miss class?
- Do we want participants who miss class to send their thoughts on the discussion topic so we can still have their input?

Objective 4: Identify qualities of good research projects for undergraduate mentees (15 min)

ACTIVITY #5: Brainstorm Elements of Effective Research Projects (15 min)

ASK ▶ What are the elements of a good research project *in your discipline* for an undergraduate? Write the ideas generated on the board or flip chart.

NOTE ▶ Some example responses may include the following:

- The project should be possible to complete (or at least make significant progress on) in the time the undergraduate has to work on it.
- The project should be built on the mentee's prior experience and knowledge, but incorporate new learning challenges.
- The project should be aligned with the mentee's interests.

NOTE ▶ Make a point of noting the differences that come up from different disciplines.

Objective 5: Establish a good relationship with your mentee (10 min)

ACTIVITY #6: Shared Ideas for Introductory Activities to Engage Mentees

TELL ▶ (5 min) Tell participants to pair up to discuss activities for their first 30 minutes with their mentees.

DISCUSS ▶ (5 min) Each pair shares ideas with the larger group.

Postsession Assignments

TELL ▶ Tell participants to write one paragraph describing their mentee's research project (or one they would have their mentee do) and a second paragraph discussing the image of science the project conveys to the mentee. Tell them to bring the paragraph with them to the next session.

TELL ▶ Tell participants to draft a mentoring philosophy/strategy (no length requirement). They can refer to the handout "Questions to Consider When Writing a Mentoring Philosophy" for ideas.

TELL ▶ Tell everyone to read the handout "Mentoring: Learned, Not Taught."

FACILITATION GUIDE

Introductory Activities: Ways to Help Participants Get to Know One Another

1. Visual Explorer

Spread out 30 or more pictures that broadly depict phenomena related to teaching and mentoring situations. Participants choose a visual representation in response to a question or statement, such as "Choose a picture that best represents mentoring." Each participant explains their choice of picture. In addition to using Visual Explorer, pictures can also be obtained as a packet of postcards, pages from a magazine, or printed images from websites, or participants can be asked to find an image on their own and bring it in. (Adapted from Paulus, C. J., Horth, D. M., and Drath, W. H. (1999), *Visual Explorer: A Tool for Making Shared Sense of Complexity*, Center for Creative Leadership Press, http://www.ccl.org/leadership/index.aspx.)

2. Significant Mentor

Have participants think of a mentor they have had that influenced their own practices. This could be a positive or negative example. Have each person briefly share what they learned.

3. Who Are You?

Participants add fun information about themselves to the four corners of their name tags. Some examples include

Hometown

Favorite food

Favorite TV show

Hobby

Favorite kind of music

Number of people in their family (How each person defines family can be very interesting!)

4. Interviews

Participants interview the person next to them and vice versa, and then introduce one another to the larger group.

5. Truth or Lie?

Everyone tells two truths and one lie, and then the group guesses the lie for each person.

6. Memorable Moments

Each person shares something memorable about their mentoring experience (as a mentor or mentee) and their motivation for participating in research mentor training.

7. Letter Names

Each person says their name and shares characteristics that start with the first letter of their name.

FACILITATION GUIDE

8. The M&M Game

Pass around a dish of M&M candies and tell participants to take as many as they would like. Ask them to introduce themselves by sharing as many characteristics about themselves as is equal to the number of M&Ms they took from the dish.

Constructive and Destructive Group Behaviors*

Choose your single most constructive group behavior and your single most destructive group behavior from the list below. Share your choices with the members of your group so they may draw on your constructive behavior and minimize your destructive behavior as you work together.

Constructive Group Behaviors

Cooperating: Is interested in the views and perspectives of other group members and willing to adapt for the good of the group.

Clarifying: Makes issues clear for the group by listening, summarizing, and focusing discussions.

Inspiring: Enlivens the group, encourages participation and progress.

Harmonizing: Encourages group cohesion and collaboration. For example, uses humor as relief after a particularly difficult discussion.

Risk Taking: Is willing to risk possible personal loss or embarrassment for the success of the overall group or project.

Process Checking: Questions the group on process issues, such as agenda, time frames, discussion topics, decision methods, and use of information.

Destructive Group Behaviors

Dominating: Uses most of the meeting time to express personal views and opinions. Tries to take control by use of power, time, and so on.

Rushing: Encourages the group to move on before the task is complete. Gets tired of listening to others and working with the group.

Withdrawing: Removes self from discussions or decision making. Refuses to participate.

Discounting: Disregards or minimizes group or individual ideas or suggestions. Severe discounting behavior includes insults, which are often in the form of jokes.

Digressing: Rambles, tells stories, and takes group away from primary purpose.

Blocking: Impedes group progress by obstructing all ideas and suggestions ("That will never work because . . .").

* Adapted from Brunt (1993). "Facilitation Skills for Quality Improvement." *Quality Enhancement Strategies*. 1008 Fish Hatchery Road. Madison WI 53715.

PARTICIPANT MATERIALS

Questions to Consider When Writing a Mentoring Philosophy

1. Why might it be useful to your own development as a research mentor to write a mentoring philosophy?
2. Why might a mentoring philosophy be useful for your mentee(s)?
3. For whom (what audiences) will you write your mentoring philosophy?
4. If you were a member of a review committee, what topics would you look for or expect to be addressed in a mentoring philosophy?
5. What would make your mentoring philosophy stand out (positively or negatively) among the 20, or 200, a review committee may have already read?
6. How should a mentoring philosophy relate to and/or incorporate into a teaching philosophy?

Mentoring: Learned, Not Taught[*]

Identifying Challenges

by Jo Handelsman

Becoming a good mentor takes practice and reflection. Each of us tends to focus on certain aspects of mentoring, which we choose for many different reasons. Sometimes we focus on issues that were important to us as mentees, those we think are hard or uncomfortable to deal with (making us worry) or easy to handle (consequently making us feel good about our mentoring), or areas in which a mentee needs help. But few of us think about the diversity of issues that comprise the full mentoring experience, at least not when we are just starting out as mentors. By broadening our approach, and looking at mentoring in a systematic way, we can become more effective mentors more quickly than if we just confront the challenges as we stumble upon them. Some of us take decades to recognize all these facets of mentoring; others of us would never discover them on our own.

This chapter focuses largely on mentors of undergraduates in a research lab, but many of the same issues arise in mentoring colleagues and others outside the lab. Each of us is likely to engage in numerous relationships as mentors and mentees throughout our careers, and each relationship will be enhanced by what we learned in the previous one. Reflecting on the following areas as your mentoring relationships evolve may help you avoid some common mistakes and hasten your arrival at a mentoring style and philosophy that is your own.

Mentoring principles, not practices, are universal

Although no one can provide formulas, practices, or behaviors that work in every mentoring situation, there are some principles that should always guide mentoring relationships. It's a good idea to ask yourself periodically whether you are adhering to your own basic principles. The values that most scientists would agree are inviolate in any mentoring relationship are honesty, kindness, caring, and maintenance of high ethical and scientific standards. As you consider the differences among students and design your mentoring strategies to serve them best, examine your values.

Mentees are different . . . from each other and from us

The diversity that our students bring us sustains the vibrancy of the scientific community and of science itself. Although most of us believe this in the abstract, dealing with people who are different from us or from our mental image of the ideal student can be frustrating and baffling. Those of us who are very organized, punctual, polite, tidy, diligent, smart, socially adept, witty, verbal, creative, confident, and tenacious probably value those characteristics in ourselves. When confronted with a mentee lacking any of them, we may wonder if they are cut out to be a scientist. Moreover, cognitive styles (the ways that we learn or think about problems) are often what scientists value most highly in themselves, but cognitive styles are idiosyncratic; thus, being a good mentor necessitates accommodating a style that differs from our own.

After we have worked with a student for a few weeks or months, we may begin to see performance issues that didn't emerge immediately. Some issues are small, some global. We may find that it drives us nuts that a student likes to work from noon until midnight, whereas we prefer working

[*] This reading first appeared in Handelsman, J., Pfund, C., Miller Lauffer, S., and Pribbenow, C. M. (2005), *Entering Mentoring: A Seminar to Train a New Generation of Scientists,* Madison, WI: University of Wisconsin Press.

READING

in the early morning. Or a student may seem unable to articulate the objectives of a research project even after substantial discussion and reading. Or the student may seem unable to get a product from a chemical reaction. Or come up with an original idea. There are no simple prescriptions for what to do. The following sections offer some questions for reflection and sample situations to provoke thought about dealing with these very complex, very human mentoring challenges.

Building confidence

Probably the most important element of mentoring is learning that performance is the product of a complex interaction among innate ability, experience, confidence, education, and the nature of the performance environment. We have all had the experience of saying something eloquently and smoothly in one setting and then stuttering our way through the same words in a stressful setting. We have the *ability* to formulate the idea and express it well, but the stressful situation affects our performance.

"performance is the product of a complex interaction among innate ability, experience, confidence, education, and the nature of the performance environment"

Confidence is influenced by the messages we receive. If we are told as children that we are very smart, we develop confidence in our intelligence. In contrast, if we are told that we can't do science because we are female or a member of the wrong ethnic group, we may have lingering doubts even when we reach the highest levels of achievement. If we come from a family in which we are the first to go to college, we may feel that we just don't quite fit in when we are in the academic environment. All of these insecurities surface at the most stressful times—when things aren't going well in the lab, when we are getting ready for exams, when we receive a poor grade, when our grants aren't funded and our papers are rejected. Those are the times when a mentor can make a difference. People with stores of confidence fall back on internal reinforcement during the rough times. The voice of a parent or teacher from the past saying "you can do it" may get them through. But people who haven't received those messages may need to hear them from a trusted mentor or colleague in order to keep going.

The challenge for many of us is to not fall into the habit of measuring every student against our own strengths. Most of us have the impulse to think, "I never needed so much support or coddling, so why should I have to give it to my students?" or "Can they really make it in science with such a need for reinforcement and coaching?" But the job of a mentor is to set high standards for mentees and then help them meet those standards. One of the most satisfying parts of mentoring is the frequency with which students surprise us. So often we hear a colleague say that, although they pushed a student to be great, it was a surprise when the student actually became great. A mentor may help a student develop the skills to be an outstanding scientist, but the most important message a mentor can ever send is that they have faith that the mentee will succeed. That faith, followed by the mentee producing high-quality science, will generate confidence.

Judging aptitude—can we?

Assessing aptitude has its own set of challenges. Because of the intersection of social, psychological, experiential, and innate factors that affect our intellects and our ability to perform, it can be difficult to judge a student's ability to be a scientist. As mentors, it is our responsibility to examine the factors affecting a student's performance. Here are a few questions we should ask:

- Are my expectations reasonable for a scientist at this stage?
- Has this student had the training necessary to succeed at this task or in this environment (and could additional formal training improve their performance)?
- Does the student understand what is expected?
- Is this student disadvantaged in some way that makes the situation more difficult than it is for others?
- Is the student experiencing a stress—inside or outside the lab—that is affecting their performance?
- Might the student perform better in another environment?

Determining whether your expectations are clear and appropriate and whether a student has the necessary preparation can be accomplished through a dialogue with the student. The solutions to these issues should be agreed upon and implemented jointly. If the remedies do not result in satisfactory performance, then other actions may need to be taken.

Judging aptitude—impact of stress

People under stress cannot work at their highest potential; it may be impossible, therefore, to judge a stressed student's aptitude for science. Stress derives from many sources, some of which are obvious, some not so apparent. The tension that we experience around deadlines is perceived by and understandable to most of us. But some students experience difficulties that may be invisible to us, and possibly even to the students themselves. Chronic illness and pain, financial problems, family responsibilities such as taking care of children or aging parents, or simply being different from the people around us can cause debilitating stress.

Some stress may come from past experience with prejudice. A student may worry that others will treat him differently if they find out that his parents are migrant farm workers, that he has epilepsy, or that he considered becoming a priest before choosing science. The student may have confronted bigotry in other situations that generated these fears and made him ultrasensitive to perceived or real intolerance. The student may be encountering prejudice in the lab that you may or may not perceive. There may be cliques from which he is excluded, jokes about his "difference" that may be intended to hurt him or are inadvertently hurtful. Discrimination experienced outside the lab or even off campus might affect the student's ability to work. A person subjected to prejudice undergoes physiological changes in many different organ systems that translate into cognitive changes that influence the ability to focus, concentrate, and be creative. Even the fear or anticipation of such attitudes (known as "stereotype threat") can have crippling effects.

If you suspect your student is suffering from stress that is affecting their ability to do science, consider discussing it with them. If the student has not discussed it with you, don't make assumptions or plunge in with aggressive questioning unless you know them very well and have established a trusting relationship. Instead, you can provide an opening for the student to seize. See the sidebar for some helpful guidelines.

Questionable Questions (unless you have already developed a trusting relationship)	(Probably) Safe Openers
Are you having marital problems? Did you break up with your girlfriend?	You seem a little down these days. Is everything OK?
Are you pregnant?	You're looking tired. I hope you're feeling OK.
Are you spending too much time at the nursing home with your mother when you should be in the lab?	Is your mother recovering from the stroke? (assuming the student had confided in you about the stroke)
What's it like to be a black man in this town, anyway?	I can imagine that being black in this very white environment might be difficult at times. If you ever want to talk about it, I'm here.
It must be hard to explain what you do to your family with no college graduates!	I was at a dinner with a bunch of lawyers the other night and, wow, did I struggle to explain what our lab does. Have you found any good analogies that laypeople can relate to?
You're so attractive, you must get a lot of attention from the guys in the lab. Is it OK being the only woman on the 12th floor?	Are you comfortable in the lab? If there are ever conflicts, problems, or issues that get in the way of your work, will you please let me know what I can do to help?
Do you want to use my office during the day to pump milk while you're breastfeeding?	I can imagine that there are lots of logistical and practical issues that will arise when you have the baby. Please let me know if there is anything I can do to make things easier for you.
Getting here for your graduation must be hard for your parents on a trash collector's salary, so do you want to use some of my frequent flyer miles to get them plane tickets?	I know you are counting on your parents being here for graduation. If there is anything I can do to help with their visit, let me know.

The inappropriate questions listed in the sidebar are all well intended, but they may call attention to something that a student doesn't want singled out, causing embarrassment or awkwardness. If your students don't want to discuss their family, race, or nursing habits with you, respect that. The more appropriate questions attempt to provide an opening that the student can take or decline. These questions express caring and show that you notice them as human beings, without intruding into private places where you might not be welcome.

Judging aptitude—innate ability

Many of us are frustrated that our students don't seem as smart as we think they should be. People mature intellectually at different rates, and all the factors discussed in the previous sections can affect apparent intelligence. It is also important to look around at people who have advanced in science and notice the characteristics that got them there. Some are simply brilliant, and the sheer power of their intellects has driven their success. But most have many other attributes that contributed to their success. Most highly successful scientists are extremely hardworking, terrific managers and motivators of other people, colorful writers, and charismatic people. The fortunate (and often most successful) scientists have large doses of all these traits, but many scientists have a mixture of strengths and weaknesses. Some are poor managers, others are unimpressive writers, and, amazingly, some don't seem all that smart or creative, yet their labs turn out great work because of their ability to create a highly effective research group.

There is room for lots of different kinds of people and intellects in science. A student who frustrates you with an excruciatingly linear or earthbound style of thinking may develop into a reliable and indispensable member of a research team. A student who can't seem to keep track of details in the lab may turn out to be a terrific professor who generates big ideas and relies on lab members to deal with the details. Before you judge a student, consider the diversity of people who make up the scientific community and ask yourself whether you can see your student being a contributor to that community. And ask yourself what each of those members of the community was like when they were at your student's stage of development.

Case 1.

I had an undergraduate student in my lab who didn't seem very bright and I doubted that he would make it as a scientist. I encouraged him to move on. The next time I saw him, he was receiving an award for outstanding undergraduate research that he did in another lab. I was surprised. The next time I encountered him was when I opened a top-notch journal and saw a paper with him as first author. I was impressed. Next I heard, he had received his PhD and was considered to be a hot prospect on the job market.

A couple of years later, I had a graduate student who was incredibly bright and a wonderful person, but wasn't getting anything done. I had tried all my mentoring tricks and then borrowed some methods from others. In a fit of frustration, I encouraged the student to take a break from the lab and think about what to do next. While she was taking her break, she received an offer to complete her PhD in another lab. She did, published a number of highly regarded papers, landed a great postdoc, and is now a well-funded faculty member at a major research university.

These experiences have made me realize the power of the "match." The student, the lab, and the advisor have to be well matched, and all of it has to come together at the right time in the student's life. I can't be a good adviser to all students, and where I fail, someone else may succeed. It reminds me to be humble about mentoring, not to judge students, and never predict what they *can't* do. Happily, they will surprise you!

Fairness: monitor prejudices and assumptions

Most of us harbor unconscious biases about other people that we apply to our evaluation of them. Few of us intend to be prejudiced, but culture and history shape us in ways that we don't recognize. Experiments show that people evaluate the quality of work differently if they are told that a man or a woman, a black or a white person performed the work (see "Benefits and Challenges of Diversity" in Chapter 5 for a detailed discussion of this research). We can't escape our culture and history, but we can try to hold ourselves to high standards of fairness and to challenge our own decisions. Regularly ask yourself if you would have reacted the same way to a behavior, a seminar, a piece of writing, or an idea if it had been presented by someone of a different gender or race. When you evaluate people, make sure you are holding them all to the same standards. When you write letters of recommendation, check your language and content and make sure that you are not introducing subtle bias with the words you use or topics you discuss (see Chapter 5 for research on letters of recommendation for men and women).

Changing behavior

When we discover that a student is disorganized, introverted, or chronically late, what should we do? How much do we accommodate these differences to encourage diversity in our research community, and when does accommodation become bad mentoring, hypocrisy, or a violation of the principles that we have agreed form our mentoring foundation? When is a behavior something that other students should tolerate, and when does it violate the rights of others in the lab? These distinctions are tough to make, and we are likely to arrive at conclusions that differ from those of other mentors or even from our own judgments at other stages in our careers. Considering a few key questions may help clarify our mentoring decisions.

- Is the behavior creating an unsafe environment for the mentee or others in the lab?
- Is the behavior negatively affecting the productivity or comfort of others in the lab?
- Will the mentee be more effective, productive, or appreciated in the lab if the behavior or characteristic is modified?
- Is the behavior or characteristic sufficiently annoying to you that it interferes with your ability to work with the mentee?

Choose your battles carefully. If your answers to the questions are all "no," you may want to let the situation go. Sloppiness that creates a fire hazard or leads to poor data record keeping must be corrected, but perhaps a desk strewn with papers, however irritating, can be ignored. A student who is introverted might be accommodated, but a student who is excessively talkative or boisterous and interfering with others' work needs to modify the behavior.

So, if a behavior needs to be changed, what's a mentor to do? If you are lucky, simply making the mentee aware of it may solve the problem. It helps to be directive about the type of change needed and why it is necessary. It is useful to lay out the problem that you are trying to solve and then ask the mentee to participate in developing the solution. If this doesn't work, you may need to use stronger language and eventually use sanctions to achieve the needed change. See the sidebar for suggestions on how to approach such matters.

Less effective	More effective
"Clean up your bench!"	"I'm concerned that the condition of your bench is creating a fire hazard. I'm sure you don't want to put the safety of the lab at risk, so what can we do to fix the situation?"
"Be on time to lab meetings from now on."	"You know, when you come into the lab meeting 15 minutes late, it's disruptive to the group and makes the person talking feel that their work isn't important to you. Is there some conflict in your schedule that I don't know about, or do you think you can be on time in the future?"
"You'll never get anywhere in science if you don't dig in and stick with problems until you solve them."	"You seem to be giving up on solving this problem. I want to help you learn how to see problems through to their solutions, so what can I do to help? I want you to work on this because problem solving is going to be important throughout your career."

Some behavior issues raise the questions of personal rights. Is it OK to rule that your students aren't allowed to wear headphones in the lab? That they dress a certain way? That they not put up posters or sayings that are offensive to others? That they aren't allowed to discuss politics or religion in the lab in ways that make some members uncomfortable? That they not make sexist or racist jokes? And whose definition of sexist and racist do we use? How do we balance overall lab happiness with the rights and needs of individuals?

Case 2.

Some issues are stickier than others. I once had a student who would come into the lab every Monday and loudly discuss his sexual exploits of the weekend. People in the lab—men and women—dreaded coming in on Mondays and were intensely uncomfortable during his discourses. No one in the group wanted to deal with it, and most of them were too embarrassed to even mention it to me. Finally, my trusted technician shared with me her intention to quit if this student didn't graduate very soon. I was faced with the challenge of telling the student that we all need to be sensitive to others in the lab and there might be people who didn't want to hear about his sex life. I was uncomfortable with the conversation for a lot of reasons. First, I'm not used to talking to my students about their sex lives. Second, I was concerned that the student would be hurt and embarrassed that others in the lab had talked to me about his behavior and I didn't want to create a new problem in the process of solving the original one. Third, the student was gay and I didn't want him to think that his behavior was offensive because of this. I wanted him to appreciate that any discussion of sexual experience—straight or gay—was simply inappropriate for the open lab environment. But the student had never told me that he was gay, so I felt it was a further violation of his relationship with other lab members to indicate that I knew he was gay. The discussion did not go well because we were both so uncomfortable with the subject and I had trouble being as blunt as I should have been. The behavior didn't change. The student finished his thesis and defended it. At the defense, one of the committee members suggested that the student do more experiments, and I detected the beginnings of a groundswell of support for his point of view. I blurted out that if this student stayed one more day in my lab, my wonderful technician would quit, so if he had to do more experiments, could he do them in one of their labs? In the end, everyone signed off on the thesis, the student graduated, and I never published the last chapter of the student's thesis because more experiments were needed to finish the story. I felt that I had weighed lab harmony against academic and scientific standards and have never been happy with how I handled the whole situation.

Deciding what to do about problematic behavior may be one of the most annoying parts of being a mentor or lab leader. Many of us just wish everyone would know how to behave, get along, and get on with the science that we are here to do. Unfortunately, behavioral issues can prevent the science from getting done, and they don't just go away. Not dealing with some problems is unfair to the mentee, who deserves to know how he or she affects others, but the behavior must be addressed in a sensitive way to prevent embarrassment and animosity. Another question is, who should handle it? If you are a graduate student responsible for an undergraduate researcher, should you take care of the problem or ask your adviser to deal with it? If you are a lab leader, should you always deal with problems directly, or is it sometimes appropriate to ask a member of the lab to tackle the problem diplomatically? These questions have to be answered in context and usually based on discussion with the other person who shares responsibility for the mentee.

Case 3.

As a graduate student, I worked in a very crowded lab that hosted two students from Puerto Rico one summer. The students were great—they worked hard, got interesting results, were fun to be around, and fit into the group really well. The problem was that they spoke Spanish to each other all day long—and I mean *all day*. For eight or nine hours every day, I listened to this loud, rapid talking that I couldn't understand. Finally, one day I blew. I said in a not-very-friendly tone of voice that I'd really appreciate it if they would stop talking because I couldn't get any work done. Afterward, I felt really bad and apologized to them. I brought the issue to my mentoring class and was surprised by the length of the discussion that resulted. People were really torn about whether it is OK to require everyone to speak in English and whether asking people not to talk in the lab is a violation of their rights. Our class happened to be visited that day by a Norwegian professor, and we asked her what her lab policy is. She said everyone in her lab is required to speak in Norwegian. That made us all quiet because we could imagine how hard it would be for us to speak Norwegian all day long.

Every mentoring relationship is different

Each person we mentor has their own unique set of needs and areas for growth. Use the beginning of the mentoring relationship to get to know your mentee and begin to experiment with ways of interacting. Does your mentee ask a lot of questions, or do they need to be encouraged to ask more? Does your mentee respond well to direct criticism, or do they need to be gently led to alternative answers or ways of doing things? In what areas do you think you can help your mentee the most—developing confidence, independence, and communication skills? Learning lab techniques and rigorous thinking? Improving interpersonal interactions? Does your mentee demand more time than you can or want to give, or do they need encouragement to seek you out more often? Mentoring relationships are as diverse as people, and they change over time. Monitor the relationship and make sure your mentoring style and habits are keeping up with the development of your mentee and the mentoring relationship.

As you assess progress in your mentoring relationship:

- Find your style—mentoring is personal and idiosyncratic.
- Communicate directly.
- Emphasize in your mentoring the aspects of science that are the most important—ethics, rigorous analytical thinking, risk taking, creativity, and people.
- Be positive. Remember that people learn to identify quality by having both the positive and the negative pointed out.
- Celebrate the differences among students.
- You are shaping the next generation—what do you want that generation to be?

An Important Mentor

One of my most important mentors was Howard Temin. He had received the Nobel Prize a few years before I met him, but I didn't discover that until I had known him for a while and I never would have guessed, because he was so modest. Many aspects of science were far more important to Howard than his fame and recognition. One of those was young people. When he believed in a young scientist, he let them know it. As a graduate student, I served with Howard on a panel about the impact of industrial research on the university. It was the first time I had addressed a room full of hundreds of people, including the press. My heart was pounding and my voice quavered throughout my opening remarks. I felt flustered and out of place. When I finished, Howard leaned over and whispered, "Nice job!" and flashed me the famous Temin smile. I have no idea whether I did a nice job or not, but his support made me feel that I had contributed something worthy and that I belonged in the discussion. I participated in the rest of the discussion with a steady voice.

When I was an assistant professor, I only saw Howard occasionally, but every time was memorable. One of the critical things he did for me—and for many other scientists—was to support risky research when no one else would. Grant panels sneered at my ideas (one called them "outlandish") and shook my faith in my abilities. Howard always reminded young scientists that virologists had resisted his ideas too, and reviews of his seminal paper describing the discovery of reverse transcriptase criticized the quality of the experiments and recommended that the paper be rejected! Howard was steadfast in his insistence that good scientists follow their instincts. When my outlandish idea turned out to be right, I paid a silent tribute to Howard Temin.

Howard showed support in many ways, some of them small but enormously meaningful. He was always interested in my work and often attended my seminars. When he was dying of cancer, his wife, Rayla, also a scientist, went home each day to make lunch for him. During that time, I gave a noon seminar on teaching that Rayla mentioned to Howard. When he heard who was giving the seminar, he told Rayla to attend it and that he would manage by himself that day. That was the last gift Howard gave me as a mentor before he died, and it will always live with me as the most important because it embodied everything I loved about Howard: he was selfless, generous, caring, and supportive.

At Howard's memorial service, students and colleagues spoke about how they benefited, as I had, from his enormous heart and the support that gave them the fortitude to take risks and fight difficult battles. Each of us who was touched by Howard knows that he left the world a magnificent body of science, but to us, his greatest legacy is held closely by the people who were lucky enough to have been changed by his great spirit.

READING

CHAPTER 2

Aligning Expectations

Introduction

One critical element of an effective mentor-mentee relationship is a shared understanding of what each person expects from the relationship. Problems between mentors and mentees often arise from misunderstandings about expectations. Importantly, expectations change over time, so frequent reflection and clear communication about expectations are needed on a regular basis.

Learning Objectives

Mentors will have the knowledge and skills to

1. **Design and communicate clear goals for the research project**
2. **Listen to and consider the expectations of their mentee in the mentoring relationship**
3. **Consider how personal and professional differences may impact expectations**
4. **Clearly communicate expectations for the mentoring relationship**
5. **Align mentee and mentor expectations**

Overview of Activities for the "Aligning Expectations" Session

	Learning Objectives	Core Activities
1	Design and communicate clear goals for the research project	Share mentee project descriptions (Activity #1)
2	Listen to and consider the expectations of the mentee in the mentoring relationship	Read and discuss the case study (Activity #2)
3	Consider how personal and professional differences may impact expectations	Discuss additional guiding questions to the case study
4	Clearly communicate expectations for the mentoring relationship	Review mentoring compacts (Activity #3)
5	Align mentee and mentor expectations	Discuss expectations with mentee or draft compact (Post-session Assignment)

FACILITATION GUIDE

Recommended Session for Aligning Expectations (60 minutes)

Materials Needed for the Session:

- Table tents
- Chalkboard, whiteboard, or flip chart
- Handouts:
  - Copies of introduction and learning objectives for Aligning Expectations (page 27)
  - Copies of the case study, "The Sulky Undergraduate" (page 31)
  - Copies of example mentor-mentee compacts (pages 32–40)
  - Copies of Mentoring Tools (pages 41–44)

Introductions (10 min)

ASK ▶ Ask participants to remind everyone in the group who they are and ask them to share one word or phrase that describes the typical undergraduate research mentoring experience in their discipline.

TELL ▶ Review the introduction and learning objectives for the session.

Objective 1: Design and communicate clear goals for the research project (15 min)

ACTIVITY #1: Sharing Research Project Descriptions

ASK ▶ Ask participants to pair off and read one another's project descriptions (or verbally share the project description if they did not bring the written assignment).

DISCUSS ▶ (10 min) Have the partners talk about the image of research conveyed by their descriptions and how the role of the mentee could be perceived after reading the descriptions.

DISCUSS ▶ (5 min) Discuss with the entire group how their descriptions can be improved to convey the excitement of research as well as to reflect expectations for the mentee .

Objectives 2 and 3: Listen to and consider the expectations of their mentee in the mentoring relationship and consider how personal and professional differences may impact expectations (15 min)

ACTIVITY #2: Case Study: The Sulky Undergraduate

READ ▶ Distribute the case study (see page 31) and ask one participant to read the case aloud.

DISCUSS ▶ (15 min) Have a group discussion about the study. You may want to record the ideas generated in this discussion on the whiteboard or flip chart. Use the guiding questions following the case study to facilitate discussion. Additional questions are as follows:

- How do you establish and communicate your expectations to your mentee?
- When choosing a project for your mentee, how do you weigh the mentee's interest with the immediate needs of the research PI or group?
- As an adviser or mentor, what should you do if a mentee does not like the project?
- How do you find out your mentee's expectations of you and of the research experience?
- How can you make sure your expectations take into account a mentee's individual learning style, background, and abilities?
- How do you assess your mentee's skills so you can choose an appropriate project?

Objective 4: Clearly communicate expectations for the mentoring relationship (15 min)

ACTIVITY #3: Reviewing Mentor-Mentee Compacts

ASK ▶ Ask the participants if any of them use mentor-mentee compacts. If so, what has their experience been in using them?

NOTE ▶ We use the term *compacts* in this curriculum, but others refer to these expectations documents as *contracts*. Both are agreements between two parties, and we use the terms interchangeably. However, contracts are legally binding and compacts are not.

TELL ▶ The sample compacts provided on pages 32–40 include two that have been used primarily with undergraduate students and one used with graduate students. Introduce their use to the participants in the following way:

- It is likely that some of the items will resonate with you and others will not. The goal today is to identify those elements that you would include in your own compact and note additional items you would like to incorporate later. If the idea of filling out a compact and having your mentees do the same is off-putting or too formal for you, consider the other ways compacts can be useful. The act of writing a compact is a simple way to organize your thoughts and values as a mentor. Even if you never share the compact itself, it may be a useful tool to prepare you for a meeting with a mentee. The compact may also provide useful information to or guide a discussion with your mentee. It is useful to work through the process of designing a compact, even if you don't use it as a formal agreement.

Additional compacts can be downloaded from https://mentoringresources.ictr.wisc.edu/ExampleMentoringCompacts. Notice the differences in expectations across career stage.

TELL ► Remind mentors that they may create a template document listing expectations that can be used to initiate a discussion of this topic with mentees, but the essential component is the *process* of sharing goals and expectations and arriving at a common understanding. Individual Development Plans, like those included in the "Promoting Professional Development" session (Chapter 3), can be utilized in concert with the mentor's expectations template to tailor a holistic plan for each mentee.

TELL ► (5 min) Tell participants to review the sample compacts individually and circle or highlight the items they would like to include in their own compact or topics they want to make certain to discuss with their mentee.

ASK ► (5 min) Pose the following questions to the entire group: What are your initial thoughts about using a mentor-mentee compact? What are your initial thoughts about the examples provided?

DISCUSS ► (5 min) Have the mentors pair off to discuss items chosen for their own compacts.

Objective 5: Align mentee and mentor expectations (5 min)

Postsession Assignments

TELL ► Instruct participants to interview their mentee and write a paragraph that describes their mentee.

TELL ► Tell participants to develop or adapt a compact/contract to use with a current or future mentee.

Mentoring Tools[*]

The following mentoring tools are located at the end of this chapter:

- Research Experience Expectations
- Research Experience Reflections
- Letter of Recommendation
- Roles for Your Research Mentor

These tools are intended to help mentees identify, and subsequently share with their mentors, the expectations they have for the mentoring relationship.

---

[*] These mentoring tools are adapted from Branchaw, J. L., Pfund, C., and Rediske, R. (2010), *Entering Research: A Facilitator's Manual: Workshops for Students Beginning Research in Science*, W.H. Freeman & Company.

Case Study: The Sulky Undergraduate

I mentored an undergraduate student who came from another university for the summer. I explained the project to him and taught him some basic techniques and approaches needed for the project. Because my professor and I did not think he had sufficient background for a more complicated project, we chose to have him work on a more basic one.

He was very quiet for the first 10 days of the project, and then he went to my adviser and complained about the project. He said he wanted a project "like Mark's." Mark was a student with a strong disciplinary background and his project was much more advanced. My adviser insisted that my mentee keep the project I had designed for him, but the student became sulky. As the summer went on and he didn't get much of his work done, I began to wonder if he understood what we were doing or even cared about it.

Guiding Questions for Discussion:

1. What are the main themes raised in this case study?

2. What kind of conversations regarding expectations might have been helpful early in this relationship?

3. What kind of conversations would be helpful once the student asked for a different project? Who should be involved in these conversations?

Examples of Mentor-Mentee Compacts

1. Undergraduate Student Mentee Examples

Mentor-Mentee Contract (from *Entering Research: A Facilitator's Manual*)

Expectations for Undergraduate Mentees (from Ashley Shade, University of Wisconsin–Madison)

2. Graduate Student Mentee Example (from Professor Trina McMahon, University of Wisconsin–Madison)

Note: All the examples presented here are designed for one-on-one mentor-mentee relationships. If a mentee has multiple mentors, then the mentee may have individual compact agreements with each mentor or create one compact to which everyone agrees.

Undergraduate Mentee Contract[*]

Undergraduate Mentee: _____

Graduate or Postdoc Mentor: _____

This contract outlines the parameters of our work together on this research project.

1. Our major goals are:

    A. research project goals _____

    B. mentee's personal and/or professional goals _____

    C. mentor's personal and/or professional goals _____

2. Our shared vision of success in this research project is:

3. We agree to work together on this project for at least _____ semesters.

4. The mentee will work at least _____ hours per week on the project during the academic year, and _____ hours per week in the summer.

    The mentee will propose his/her weekly schedule to the mentor by the _____ week of the semester.

    If the mentee must deviate from this schedule (e.g., to study for an upcoming exam), he or she will communicate this to the mentor at least _____ (weeks / days / hours) before the change occurs.

5. On a daily basis, our primary means of communication will be through (circle):
   face-to-face / phone / email / instant messaging / _____

6. We will meet one-on-one to discuss our progress on the project and to reaffirm or revise our goals for at least _____ minutes _____ time(s) per month.

    It will be the (mentee's / mentor's) responsibility to schedule these meetings. (circle)

    In preparation for these meetings, the mentee will:

    In preparation for these meetings, the mentor will:

_____

* Adapted from Branchaw, J. L., Pfund, C., and Rediske, R. (2010), *Entering Research: A Facilitator's Manual: Workshops for Students Beginning Research in Science*, W.H. Freeman & Company.

PARTICIPANT MATERIALS

At these meetings, the mentor will provide feedback on the mentee's performance and specific suggestions for how to improve or progress to the next level of responsibility through (circle):

a. a written evaluation   b. a verbal evaluation   c. other: _____

7. The mentor will train the mentee on new techniques and procedures using the following (e.g., written directions, hands-on demonstration, verbal direction as mentee does procedure, etc.):

8. If the mentee gets stuck while working on the project (e.g., has questions or needs help with a technique or data analysis), the procedure to follow will be:

9. The standard operating procedures for working in our research group, which all group members must follow and the mentee agrees to follow, include (e.g., wash your own glassware, attend weekly lab meeting, reorder supplies when you use the last of something, etc.):

10. Other issues not addressed above that are important to our work together:

By signing below, we agree to these goals, expectations, and working parameters for this research project.

Mentee's signature: _____ Date: _____

Mentor's signature: _____ Date: _____

Professor's signature: _____ Date: _____

Expectations for Undergraduate Mentees[*]

1. **Send me weekly email updates on Fridays by 5 p.m.**, describing briefly what you've been working on, what you plan to do the following week, and any questions or troubles you had. Important things to include: project you've worked on, broken equipment, storage/equipment conflicts, if your data look weird.

2. **Attend lab meeting.** The entire lab assembles approximately once a week to discuss our research. Generally, the person leading lab meeting will distribute reading materials in advance. You should read these materials and come prepared to participate actively in the discussion.

3. **Be organized.** There is a lot of overlap in projects, and it is essential that you keep track of all the samples in the way that I specify. This includes updating the data spreadsheets and lab notebooks immediately.

4. **Read background information and protocols about our projects, and about our lab's research.** This includes the protocol handout, the wiki, and related journal articles from the lab that I've suggested. I'd love to discuss any journal article or protocol, so just say the word and we'll grab some coffee and chat.

5. **Be consistent with your lab schedule.** Email/call me if you are going to be very late or unable to make your scheduled lab time.

6. **Be independent.** I am periodically away, and I expect you to get things done well without me. Ask questions when I am around, but don't be afraid to try to do detective work on your own if I am not. We have a helpful, experienced lab, so know that folks other than me may be excellent resources.

7. **Respect the lab area and your colleagues.** Keep it neat and ask if you have questions on equipment use, cleaning, etc. It is very important that you tell me if a piece of equipment breaks. Do not be worried that I will be angry. These things happen all the time in labs, and the important thing is that I know it is broken and can arrange to have it fixed.

8. **Let me know if you need anything from me as a mentor, or if you have questions.** Be up-front and I will do the same.

9. **I have an "open door" policy.** Let me know if you are having troubles or concerns that you want to talk about with me, work-related or not. My phone number is XXXXXX.

* From Ashley Shade, University of Wisconsin–Madison research mentor

PARTICIPANT MATERIALS

Graduate Mentee Contract[*]

The broad goals of my research program

As part of my job as a professor, I am expected to write grants and initiate research that will make tangible contributions to science, the academic community, and society. You will be helping me carry out this research. It is imperative that we carry out good scientific method, and conduct ourselves in an ethical way. We must always keep in mind that the ultimate goal of our research is publication in scientific journals. Dissemination of the knowledge we gain is critical to the advancement of our field. I also value outreach and informal science education, both in the classroom and while engaging with the public. I expect you to participate in this component of our lab mission while you are part of the lab group.

What I expect from you

Another part of my job as a professor is to train and advise students. I must contribute to your professional development and progress in your degree. I will help you set goals and hopefully achieve them. However, I cannot do the work for you. In general, I expect you to

- Learn how to plan, design, and conduct high-quality scientific research
- Learn how to present and document your scientific findings
- Be honest, ethical, and enthusiastic
- Be engaged within the research group and at least two programs on campus
- Treat your lab mates, lab funds, equipment, and microbes with respect
- Take advantage of professional development opportunities
- Obtain your degree
- Work hard—don't give up!

You will take ownership over your educational experience

- **Acknowledge that you have the primary responsibility for the successful completion of your degree.** This includes commitment to your work in classrooms and the laboratory. You should maintain a high level of professionalism, self-motivation, engagement, scientific curiosity, and ethical standards.
- **Ensure that you meet regularly with me and provide me with updates on the progress and results of your activities and experiments.** Make sure that you also use this time to communicate new ideas that you have about your work and challenges that you are facing. Remember: I cannot address or advise about issues that you do not bring to my attention.
- **Be knowledgeable of the policies, deadlines, and requirements of the graduate program, the graduate school, and the university.** Comply with all institutional policies, including academic program milestones, laboratory practices, and rules related to chemical safety, biosafety, and fieldwork.
- **Actively cultivate your professional development.** UW–Madison has outstanding resources in place to support professional development for students. I expect you to take full advantage of these resources, since part of becoming a successful engineer or scientist involves more than just doing academic research. You are expected to make continued progress in your development as a teacher, as an ambassador to the general public representing the university and your discipline, with respect to your networking skills, and as an engaged member of broader professional organizations. The graduate school has a regular seminar series related to professional

---

[*] From Professor Trina McMahon, University of Wisconsin–Madison

development. The Delta program offers formalized training in the integration of research, teaching, and learning. All graduate degree programs require attendance at a weekly seminar. Various organizations on campus engage in science outreach and informal education activities. Attendance at conferences and workshops will also provide professional development opportunities. When you attend a conference, I expect you to seek out these opportunities to make the most of your attendance. You should become a member of one or more professional societies, such as the Water Environment Federation, the American Society for Microbiology, or the American Society for Limnology and Oceanography.

You will be a team player

- **Attend and actively participate in all group meetings, as well as seminars that are part of your educational program.** Participation in group meetings does not mean only presenting your own work, but providing support to others in the lab through shared insight. You should refrain from using your computer, Blackberry, or iPhone during research meetings. Even if you are using the device to augment the discussion, it is disrespectful to the larger group to have your attention distracted by the device. Do your part to create a climate of engagement and mutual respect.
- **Strive to be the very best lab citizen.** Take part in shared laboratory responsibilities and use laboratory resources carefully and frugally. Maintain a safe and clean laboratory space where data and research participant confidentiality are protected. Be respectful to, tolerant of, and work collegially with all laboratory colleagues: respect individual differences in values, personalities, work styles, and theoretical perspectives.
- **Be a good collaborator.** Engage in collaborations within and beyond our lab group. Collaborations are more than just publishing papers together. They demand effective and frequent communication, mutual respect, trust, and shared goals. Effective collaboration is an extremely important component of the mission of our lab.
- **Leave no trace.** As part of our collaborations with the Center for Limnology and other research groups, you will often be using equipment that does not belong to our lab. I ask that you respect this equipment and treat it even more carefully than our own equipment. Always return it as soon as possible in the same condition you found it. If something breaks, tell me right away so that we can arrange to fix or replace it. Don't panic over broken equipment. Mistakes happen. But it is not acceptable to return something broken or damaged without taking the steps necessary to fix it.
- **Acknowledge the efforts of collaborators.** This includes other members of the lab as well as those outside the lab.

You will develop strong research skills

- **Take advantage of your opportunity to work at a world-class university by developing and refining stellar research skills.** I expect that you will learn how to plan, design, and conduct high-quality scientific research.
- **Challenge yourself by presenting your work at meetings and seminars as early as you can and by preparing scientific articles that effectively present your work to others in the field.** The "currency" in science is published papers: they drive a lot of what we do. And because our lab is supported by taxpayer dollars, we have an obligation to complete and disseminate our findings. I will push you to publish your research as you move through your training program, not only at the end. Students pursuing a master's degree will be expected to author or make major contributions to at least one journal paper submission. Students pursuing a doctoral degree will be expected to be lead author on at least two journal paper submissions, preferably three or four.
- **Keep up with the literature so that you can have a hand in guiding your own research.** Block at least 1 hour per week to peruse current tables of contents for journals or do literature searches. Participate in journal clubs. Better yet, organize one!

- **Maintain detailed, organized, and accurate laboratory records.** Be aware that your notes, records, and all tangible research data are my property as the lab director. When you leave the lab, I encourage you to take copies of your data with you. But one full set of all data must stay in the lab, with appropriate and accessible documentation. Regularly back up your computer data to the server (see the wiki for more instructions).
- **Be responsive to advice and constructive criticism.** The feedback you get from me, your colleagues, your committee members, and your course instructors is intended to improve your scientific work.

You will work to meet deadlines

- **Strive to meet deadlines: this is the only way to manage your progress.** Deadlines can be managed in a number of ways, but I expect you to do your best to maintain these goals. We will establish mutually agreed upon deadlines for each phase of your work during one-on-one meetings at the beginning of each term. For graduate students, there is to be a balance between time spent in class and time spent on research and perhaps on outreach or teaching. As long as you are meeting expectations, you can largely set your own schedule. It is your responsibility to talk with me if you are having difficulty completing your work, and I will consider your progress unsatisfactory if I need to follow up with you about completion of your lab or coursework.
- **Be mindful of the constraints on my time.** When we set a deadline, I will block off time to read and respond to your work. If I do not receive your materials, I will move your project to the end of my queue. Allow a minimum of 1 week prior to submission deadlines for me to read and respond to short materials, such as conference abstracts, and 3 weeks for me to work on manuscripts or grant proposals. Please do not assume I can read materials within a day or two, especially when I am traveling.

You will communicate clearly

- **Remember that all of us are "new" at various points in our careers.** If you feel uncertain, overwhelmed, or want additional support, please overtly ask for it. I welcome these conversations and view them as necessary.
- **Let me know the style of communication or schedule of meetings that you prefer.** If there is something about my mentoring style that is proving difficult for you, please tell me so that you give me an opportunity to find an approach that works for you. No single style works for everyone; no one style is expected to work all the time. Do not cancel meetings with me if you feel that you have not made adequate progress on your research; these might be the most critical times to meet with a mentor.
- **Be prompt.** Respond promptly (in most cases, within 48 hours) to emails from anyone in our lab group and show up on time and prepared for meetings. If you need time to gather information in response to an email, please acknowledge receipt of the message and indicate when you will be able to provide the requested information.
- **Discuss policies on work hours, sick leave, and vacation with me directly.** Consult with me and notify fellow lab members in advance of any planned absences. Graduate students can expect to work an average of 50 hours per week in the lab; postdocs and staff at least 40 hours per week. I expect that most lab members will not exceed 2 weeks of personal travel away from the lab in any given year. Most research participants are available during university holidays, so all travel plans, even at the major holidays, must be approved by me before any firm plans are made. I believe that work-life balance and vacation time are essential for creative thinking and good health and encourage you to take regular vacations. Be aware, however, that there will necessarily be epochs—especially early in your training—when more effort will need to be devoted to work and it may not be ideal to schedule time away. This includes the field season, for students/postdocs working on the lakes.
- **Discuss policies on authorship and attendance at professional meetings with me before beginning any projects to ensure that we are in agreement.** I expect you to submit relevant research results in a timely manner. Barring unusual circumstances, it is my policy that students are first author on all work for which they took the lead on data collection and preparation of the initial draft of the manuscript.

- **Help other students with their projects and mentor/train other students.** This is a valuable experience! Undergraduates working in the lab should be encouraged to contribute to the writing of manuscripts. If you wish to add other individuals as authors to your papers, please discuss this with me early on and before discussing the situation with the potential coauthors.

What you should expect from me

- **I will work tirelessly** for the good of the lab group; the success of every member of our group is my top priority, no matter their personal strengths and weaknesses, or career goals.
- **I will be available for regular meetings and informal conversations.** My busy schedule requires that we plan in advance for meetings to discuss your research and any professional or personal concerns you have. Although I will try to be available as much as possible for "drop-in business," keep in mind that I am often running to teach a class or to a faculty meeting and will have limited time.
- **I will help you navigate your graduate program of study.** As stated previously, you are responsible for keeping up with deadlines and being knowledgeable about requirements for your specific program. However, I am available to help interpret these requirements, select appropriate coursework, and select committee members for your oral exams.
- **I will discuss data ownership and authorship policies regarding papers with you.** These can create unnecessary conflict within the lab and among collaborators. It is important that we communicate openly and regularly about them. Do not hesitate to voice concerns when you have them.
- **I will be your advocate.** If you have a problem, come and see me. I will do my best to help you solve it.
- **I am committed to mentoring you, even after you leave my lab.** I am committed to your education and training while you are in my lab, and to advising and guiding your career development—to the degree you wish—long after you leave. I will provide honest letters of evaluation for you when you request them.
- **I will lead by example and facilitate your training in complementary skills needed to be a successful scientist, such as oral and written communication, grant writing, lab management, mentoring, and scientific professionalism.** I will encourage you to seek opportunities in teaching, even if not required for your degree program. I will also strongly encourage you to gain practice in mentoring undergraduate and/or high school students, and to seek formal training in this activity through the Delta program.
- **I will encourage you to attend scientific/professional meetings and will make an effort to fund such activities.** I will not be able to cover all requests, but you can generally expect to attend at least one major conference per year, when you have material to present. Please use conferences as an opportunity to further your education, and not as a vacation. If you register for a conference, I expect you to attend the scientific sessions and participate in conference activities during the time you are there. Travel fellowships are available through the environmental engineering program, the Bacteriology Department, and the university if grant money is not available. I will help you identify and apply for these opportunities.
- **I will strive to be supportive, equitable, accessible, encouraging, and respectful. I will try my best to understand your unique situation, and mentor you accordingly.** I am mindful that each student comes from a different background and has different professional goals. It will help if you keep me informed about your experiences and remember that graduate school is a job with very high expectations. I view my role as fostering your professional confidence and encouraging your critical thinking, skepticism, and creativity. If my attempts to do this are not effective for you, I am open to talking with you about other ways to achieve these goals.

Yearly evaluation

Each year we will sit down to discuss progress and goals. At that time, you should be sure to tell me if you are unhappy with any aspect of your experience as a graduate student here. Remember that I am your advocate, as well as your adviser. I will be able to help you with any problems you might have with other students, professors, or staff.

Similarly, we should discuss any concerns that you have with respect to my role as your adviser. If you feel that you need more guidance, tell me. If you feel that I am interfering too much with your work, tell me. If you would like to meet with me more often, tell me. At the same time, I will tell you if I am satisfied with your progress, and if I think you are on track to graduate by your target date. It will be my responsibility to explain to you any deficiencies, so that you can take steps to fix them. This will be a good time for us to take care of any issues before they become major problems.

Mentoring Tool

Research Experience Expectations[*]

> **Objective:** Students will articulate their motivations and goals for doing research, what they bring to the experience, and what they aim to learn from the experience.

1. Why do you want to do research?

2. What are your academic and personal goals for your research experience?

3. What values, experiences, and/or perspectives will you bring to your research team?

4. What is your greatest concern about doing research?

5. What most excites you about doing research?

PARTICIPANT MATERIALS

---

[*] Adapted from Branchaw, J. L., Pfund, C., and Rediske, R. (2010), *Entering Research: A Facilitator's Manual: Workshops for Students Beginning Research in Science*, W.H. Freeman & Company.

Mentoring Tool

PARTICIPANT MATERIALS

Research Experience Reflections[*]

> **Objective:** Students will reflect on what they learned and the goals they achieved during their research experience.

1. Was your research experience what you expected it to be? Why or why not?

2. What academic and personal goals did you achieve in your research experience? How do they compare to the goals you outlined at the beginning of your experience?

3. What values, experiences, and/or perspectives did you contribute to your research team? Were you able to contribute in ways that you did not predict? How?

4. How did you overcome your greatest concern about doing research? What was the most challenging aspect of your research experience?

5. What was the best part about your research experience? Are you planning to continue doing research? Why or why not?

---

[*] Adapted from Branchaw, J. L., Pfund, C., and Rediske, R. (2010), *Entering Research: A Facilitator's Manual: Workshops for Students Beginning Research in Science*, W.H. Freeman & Company.

Mentoring Tool

Letter of Recommendation[*]

> **Objective:** Students will consider what they expect their mentor(s) to say in a letter of recommendation about them and reflect on whether their behavior and performance support these expectations.

One of the benefits of doing research is that you will get to know your mentor well, and he or she will be able to write a detailed letter of recommendation on your behalf when you move to the next stage of your academic or professional career. As you begin the research experience, reflect on the expectations and goals you established with your mentor. Consider what you would like him or her to be able to say about you at the end of the research experience and complete this draft letter of recommendation.

Date

Dear Selection Committee,

I am writing this letter in support of [*your name*], who is applying for [*job of your dreams*]. I believe [*your name*] is an excellent candidate for this position because

1.
2.
3.

Over the past few years, [*your name*] has worked in my research group in the Department of [*your department*] at the [*your campus*]. [*Your name*] is very skilled in the following areas:

1.
2.
3.

In short, I believe that [*your name*] would be a wonderful asset to your department/program/unit. I strongly recommend him/her.

Sincerely,
[*Your mentor*]

Do the things outlined in this letter of recommendation align with the goals and expectations you established with your mentor? If not, how can you adjust your goals and expectations so that you will have the opportunity to engage in activities that allow your mentor to comment on these things?

---

[*] Adapted from Branchaw, J. L., Pfund, C., and Rediske, R. (2010), *Entering Research: A Facilitator's Manual: Workshops for Students Beginning Research in Science*, W.H. Freeman & Company.

Mentoring Tool

Roles for Your Research Mentor[*]

> **Objective:** Students will consider the different roles that their research mentors play and reflect on which are most important to them.

Consider the different roles of research mentors listed below. Add other roles that may be missing from the list. Prioritize these roles according to your expectations, with #1 as the most important.

Role	Priority
Teach by example	
Train you in disciplinary research	
Improve your writing and communication skills	
Provide growth experiences	
Help build your self-confidence as a researcher	
Model and promote professional behavior	
Inspire	
Offer encouragement	
Assist with advancement of your career	
Facilitate networking with colleagues	
Help build the bridge between research and clinical work	
Other:	
Other:	
Other:	

---

[*] Adapted from Branchaw, J. L., Pfund, C., and Rediske, R. (2010), *Entering Research: A Facilitator's Manual: Workshops for Students Beginning Research in Science*, W.H. Freeman & Company.

Promoting Professional Development

Introduction

A goal of most mentoring situations is to enable the mentee to identify and achieve both academic and professional outcomes. Though learning to do disciplinary research is an important academic outcome in research mentoring relationships, there are many other outcomes that will influence a mentee's future career. Mentors should consciously consider and support their mentees to achieve these other outcomes as well.

Learning Objectives

Mentors will have the knowledge and skills to

1. **Identify the roles mentors play in the overall professional development of their mentees**
2. **Develop a strategy for guiding professional development using a written document**
3. **Initiate and sustain periodic conversations with mentees on professional goals and career development objectives and strategies**
4. **Engage in open dialogue on balancing the competing demands, needs, and interests of mentors and mentees (e.g., research productivity, coursework, creativity, independence, career preference decisions, nonresearch activities, personal development, and work-family balance)**

Overview of Activities for the "Promoting Professional Development" Session

	Learning Objectives	Core Activities
1	Identify the roles mentors play in the overall professional development of their mentees	Brainstorm a list of the roles mentors play in the professional development of their mentees beyond research, and rank them in order of importance (Activity #1)
2	Develop a strategy for guiding professional development using a written document	Review and discuss three documents that could be used as guides to create Individual Development Plans (IDPs) (Activity #2 and Postsession Assignment)
3	Initiate and sustain periodic conversations with mentees on professional goals and career development objectives and strategies	Use the written professional development plan created in Activity #2 as a guide for a conversation with their mentee about career development (Activity #3)
4	Engage in open dialogue on balancing the competing demands, needs, and interests of mentors and mentees (e.g., research productivity, coursework, creativity, independence, career preference decisions, nonresearch activities, personal development, and work-family balance)	Read and discuss the case study, "To Be or Not to Be a PhD" (Activity #4)

FACILITATION GUIDE

Recommended Session for Promoting Professional Development (60 minutes)

Materials Needed for the Session:

- Table tents
- Chalkboard, whiteboard, or flip chart
- Handouts:
  - Copies of introduction and learning objectives for Promoting Professional Development (page 45)
  - Copies of the example Individual Development Plans (pages 49–53)
  - Copies of the case study, "To Be or Not to Be a PhD" (page 54), and additional cases if desired
  - Copies of Mentoring Tool (page 55)

Introductions (10 min)

ASK ▶ Ask mentors to share one element from their draft mentoring compact/contract.

TELL ▶ Review the introduction and learning objectives for the session.

Objective 1: Identify the roles mentors play in the overall professional development of their mentees (15 min)

ACTIVITY #1: Brainstorming Mentor Roles in Professional Development

ASK ▶ (5 min) In pairs, have participants list all the roles mentors can or should play in the professional development of their mentee, beyond research training.

DISCUSS ▶ (5 min) Discuss with the entire group the roles each pair listed. You may want to record the ideas generated in this discussion on the whiteboard or flip chart, indicating which are mentioned multiple times.

NOTE ▶ Some elements of professional development include the following:
- Networking—social and professional
- Socialization to local lab culture
- Exploring future research opportunities
- Finding funding
- Time management
- Leadership skills
- Career path exploration and guidance
- Public speaking
- Research ethics
- Writing skills
- Fostering informal mentoring relationships

DISCUSS ▶ (5 min) Engage the entire group in a discussion of the following questions:
- Which of the roles on the list are the most important? Why? How might the roles change with the development and career stage of the mentee?
- Are there some roles on the list that should not be the mentor's concern? Why?
- How do you decide which roles you will play in the relationship?

Objective 2: Develop a strategy for guiding professional development using a written format (10 min)

ACTIVITY #2: Reviewing Individual Development Plans (IDPs) and Mentoring Plans

REVIEW ▶ (10 min) Have mentors review sample plans individually and make notes on them to indicate which aspects they would like to adopt for use with their own mentees. Some mentors may already use such plans and wish to share their own versions.

TELL ▶ Suggest that IDPs be used in the mentor selection process by mentees as a means of assessing fit.

NOTE ▶ Additional examples are available at https://mentoringresources.ictr.wisc.edu. Mentors may also wish to refer their mentees to http://myidp.sciencecareers.org where they can develop their IDP through a guided, online process.

Objective 3: Initiate and sustain periodic conversations with mentees on professional goals and career development objectives and strategies (10 min)

ACTIVITY #3: Using the Individual Development Plans and Mentoring Plans

DISCUSS ▶ (5 min) Have mentors pair off and share specific ways they could introduce the idea of an Individual Development Plan to their mentee and how the completed plan could be used to navigate the mentoring relationship.

DISCUSS ▶ (5 min) Discuss with the entire group. You may want to record ideas generated on the whiteboard or flip chart.

NOTE ▶ Mentoring compacts, like those included in the "Aligning Expectations" session (Chapter 2), can be utilized in concert with these IDPs to tailor a holistic plan for each mentee.

Objective 4: Engage in open dialogue on balancing the competing demands, needs, and interests of mentors and mentees (15 min)

ACTIVITY #4: Case Study: To Be or Not to Be a PhD

READ ▶ (3 min) Distribute the case study and let participants read the case individually.

DISCUSS ▶ (12 min) Engage the entire group in discussion. You may want to record the ideas generated in this discussion on the whiteboard or flip chart. Use the guiding questions following the case study. Additional questions are as follows:

- What are the responsibilities of the mentor to every mentee, regardless of career path?
- To what extent are the differing value systems of the mentor and mentee a factor in their relationship?
- Do the genders of the mentee and mentor influence your assessment of this case?
- How do issues of socialization arise in this case study? What does it look like to belong to the academic enterprise?
- How might nonresearch interests and personal goals or obligations play into a mentee's decision about career path? How might the mentor draw these factors out in discussion?

NOTE ▶ Encourage mentors to return to their mentor-mentee compact (if applicable) and include text on how both they and the mentee are expected to communicate a sudden change in the work plan due to change in career path, health issues, family issues, and so on, and how they will move forward.

Postsession Assignment

TELL ▶ Tell mentors to review IDP examples and generate or adapt one they could have their mentee complete in the future. The IDP should eventually be used to guide a conversation between mentor and mentee about professional development needs and expectations. For additional examples of IDPs visit https://mentoringresources.ictr.wisc.edu/MentorIDPTemplates.

Mentoring Tool*

The following mentoring tool is located at the end of the chapter:

- The Next Step in Your Career: Factors to Consider

This tool is intended to help mentees consider a range of factors when selecting a graduate training program, a professional school program, or a job.

---

* This tool is adapted from Branchaw, J. L., Pfund, C., and Rediske, R. (2010), *Entering Research: A Facilitator's Manual: Workshops for Students Beginning Research in Science*, W.H. Freeman & Company.

Examples of Individual Development Plans (IDPs) for Discussing Development and Career Plans

1. IDP for Undergraduates

2. General IDP Worksheet

Mentors may also wish to refer their mentees to http://myidp.sciencecareers.org where they can develop their IDP through a guided, online process.

Example 1: Individual Development Plan
for Undergraduate Researchers*

The Individual Development Plan (IDP) encourages undergraduate researchers to set goals and identify strategies that will help them to reach those goals. It is a self-tracking tool that can also be used to facilitate mentor-mentee communication and alignment of expectations.

Use the following five questions to guide development of the IDP. Annual (or more frequent) review of the plan provides opportunities to celebrate achievements, incorporate revisions, and ensure progress toward goals.

1. **What are your goals?**
   - **Ultimate goal**
     > *I will be a professor of neuroscience at a research university.*
   - **Long-term** (5–10 years)
     > *I will be a postdoctoral fellow studying the genetic basis of neurological disorders.*
   - **Intermediate-term** (2–5 years)
     > *I will earn my PhD degree in neuroscience.*
     > *I will contribute to the discovery of the genetic basis of Alzheimer's disease.*
   - **Short-term** (1–2 years)
     > *I will earn my BS degree in genetics.*
     > *I will publish my undergraduate research project in a peer-reviewed journal.*
   - **Immediate** (6 months–1 year)
     > *I will earn an A in biochemistry class.*
     > *I will learn brain slice immunohistochemical staining techniques.*
     > *I will participate in a summer research program to experience another university.*

2. **What competencies and skills will you need to successfully reach your goals?** (See the list at the end of this document for specific ideas.)
   - Disciplinary knowledge
   - Research and technical skills
   - Professional and interpersonal skills
   - Management and leadership skills

3. **What activities and experiences will you engage in to gain the competencies and skills?**
   - Taking classes
   - Tutoring, study groups
   - Technique training
   - Research experiences
   - Scientific meeting attendance
   - Professional development workshops

---

* Adapted and reprinted with permission of the Society for Advancement of Hispanics/Chicanos and Native Americans in Science and the Biomedical Postdoctoral Programs, University of Pennsylvania.

4. **How will you assess your progress in mastering these competencies and skills?**
   - Mastery of coursework
   - Mentor/instructor feedback
   - Successful experimental outcomes
   - Peer review

5. **Who will help you reach your goals and how?**
   - Teachers
   - Mentors
   - Peers
   - Family members

Goals	Competencies & Skills	Activities & Experiences	Assessment of Progress	Support People and Their Roles
Long-term 1.				
Intermediate-term 1. 2. 3.				
Short-term 1. 2. 3.				
Immediate 1. 2. 3.				

Examples of Skills

Research and Technical
- Critical reading (scientific literature)
- Experimental design
- Experimental techniques
- Computer skills
- Documentation/laboratory notebook
- Problem solving and troubleshooting
- Data and statistical analysis
- Critical analysis
- Responsible conduct of research
- Identification of new research directions and next steps

Professional and Interpersonal

- Reliability and follow-through
- Communication (oral and written)
- Writing (manuscript, grant, fellowship)
- Teaching
- Mentoring
- Collaborating and working in teams
- Giving/receiving constructive feedback
- Collegiality
- Networking

Management and Leadership

- Time management (meeting deadlines)
- Prioritizing and organizing work
- Leading and motivating others
- Research project management
- Budget management
- Supervising/managing people
- Delegating responsibility

Example 2: An IDP Worksheet (for the Next Year)

An Individual Development Plan is a professional tool that outlines objectives that you and your mentor/supervisor have identified as important for your professional development. A comprehensive review of your career goals and objectives identified at the beginning of your appointment and during your semiannual appraisal provide constructive feedback from your mentor/supervisor that can help you become an independent investigator.

Career Goals/Objectives	Educational Activities	Research Projects/Products/Dates
Goal One: Objective 1. 2. 3.		
Goal Two: Objective 1. 2. 3.		
Goal Three: Objective 1. 2. 3.		

Please describe the plan that you and your mentor have for your transition from your current position to the next position.

Additional Comments:

PARTICIPANT MATERIALS

Case Study: To Be or Not to Be a PhD

You are currently mentoring two undergraduate researchers. Both are very talented and hardworking; however, one has made it clear that his career goals do not include going to graduate school. He is interested in going to medical school. The other scholar has her heart set on pursuing her PhD and eventually becoming a professor. Lately, you find yourself spending more time giving professional development advice to the student who intends to go to graduate school. You rationalize this by saying that you are more familiar with this career path and thus have more to offer. Secretly you worry that you are writing off the other student, believing that he is not worth your time and advice if he is going to medical school.

Guiding Questions for Discussion:

1. What are the main themes raised in this case study?

2. What should the mentor do now? What value judgments are being made by the mentor?

3. How do you advise on career paths with which you do not have personal experience? How can you discuss potential career paths with your mentee in an unbiased manner?

Mentoring Tool

The Next Step in Your Career:
Factors to Consider[*]

> **Objective:** Students will consider what factors are important to them when selecting a graduate training program, a professional school program, or a job.

Cut out the boxes and rank the factors in order of importance to you.

Opportunity to work with a specific adviser, mentor, physician, or teacher	Coursework requirements
Climate of the training environment	Location
Relative value of teaching and research training	Alignment of personal goals with the offerings/opportunities
Reputation of a specific adviser, mentor, or coworker	Funding
Reputation of the department, office, program, or institution	Happiness of other graduate/medical students/coworkers in the program
Range of academic opportunities to engage in beyond graduate study	Feeling of inclusivity—seeing that there are others like you there
Type of preliminary/qualifying exam	Type of curriculum (e.g., case-based, traditional lecture, clinical)

PARTICIPANT MATERIALS

---

[*] Adapted from Branchaw, J. L., Pfund, C., and Rediske, R. (2010), *Entering Research: A Facilitator's Manual: Workshops for Students Beginning Research in Science*, W.H. Freeman & Company.

4

Maintaining Effective Communication

Introduction

Good communication is a key element of any relationship, and a mentoring relationship is no exception. As research mentors, it is not enough to say that we know good communication when we see it. Rather, it is critical that mentors reflect upon and identify characteristics of effective communication and take time to practice communication skills in the session and with their mentees.

Learning Objectives

Mentors will have the knowledge and skills to

1. **Provide constructive feedback**
2. **Use multiple strategies for improving communication (in person, at a distance, across multiple mentees, and within appropriate personal boundaries)**
3. **Engage in active listening**
4. **Communicate effectively across diverse dimensions, including varied backgrounds, disciplines, ethnicities, and positions of power**

Overview of Activities for the "Maintaining Effective Communication" Session

	Learning Objectives	Core Activities
1	Provide constructive feedback	Read and discuss the case study, "The Slob" (Activity #1)
2	Use multiple strategies for improving communication (in person, at a distance, across multiple mentees, and within appropriate personal boundaries)	Brainstorm a list of barriers to good communication and strategies for overcoming these barriers (Activity #2)
3	Engage in active listening	Mentors work in pairs, sharing current mentoring challenges and practicing active listening (Activity #3)
4	Communicate effectively across diverse dimensions, including varied backgrounds, disciplines, ethnicities, and positions of power	Mentors discuss what they learned from Activity #3, share specific strategies for improving communication between mentors and mentees, and complete Postsession Assignment

FACILITATION GUIDE

Recommended Session for Maintaining Effective Communication (60 minutes)

Materials Needed for the Session:

- Table tents
- Chalkboard, whiteboard, or flip chart
- Handouts:
  - Copies of introduction and learning objectives for Maintaining Effective Communication (page 57)
  - Copies of the case study, "The Slob" (page 61)
  - Copies of the Mentoring Tool (page 62)

Introduction (10 min)

ASK ▶ Ask mentors to write down at least one strategy or tool that they have decided to adopt to support their mentee's professional development. Share in the large group.

TELL ▶ Review the introduction and learning objectives for the session.

Objective 1: Provide constructive feedback (15 min)

ACTIVITY #1: Case Study: The Slob

ASK ▶ (2 min) Distribute the case study and ask one participant to read the case aloud.

DISCUSS ▶ (13 min) Discuss the case with the entire group. You may want to record the ideas generated in this discussion on the whiteboard or flip chart. Use the guiding questions following the case study. Here are additional questions as well:

- What is the most effective way to communicate who should be involved in dealing with problems that arise between the mentor and mentee?
- How does this situation affect the research group environment?
- How do you know if there is a problem with your mentee?

Objective 2: Use multiple strategies for improving communication (in person, at a distance, across multiple mentees, and within appropriate personal boundaries) (15 min)

ACTIVITY #2: Brainstorming Communication Strategies (15 min)

ASK ▶ Brainstorm a list of barriers to good communication.

NOTE ▶ Record the list on the whiteboard or flip chart, and then have mentors choose two or three barriers and discuss practical ways to overcome them. Mentors could generate a table such as the one presented here.

Barrier to Effective Communication	Solutions to Overcome Barrier	Indications That Communication Has Improved
Example: Lack of time to meet one-on-one	Frequent email, telecoms, or instant messaging chat time	Fewer misunderstandings and stalls in research progress

NOTE ▶ Alternatively, have the mentors create a list of all the forms of communication used by them and their mentee (face-to-face meetings, email, sticky notes, phone calls, etc.). Organize the resulting list by types of communication and assign each type to a group of two to three mentors. Each subgroup should then discuss ways each method can be improved. At the end, have the small groups report out. Record all ideas on the whiteboard or flip chart. You may want to send a compiled list to the entire group.

Objectives 3 and 4: Engage in active listening and communicate effectively across diverse dimensions (20 min)

ACTIVITY #3: Active Listening

TELL ▶ Active listening involves making a conscious effort to give your full attention to the act of listening to ensure that you understand the speaker's intent. Active listening includes maintaining eye contact with the speaker while he or she is talking, paying attention to what the speaker is saying, asking for clarification when you don't understand, and attending not only to the words but also to the feelings behind the words. More information on active listening and effective communication in general can be found at https://mentoringresources.ictr.wisc.edu.

ASK ▶ (10 min) Ask mentors to form groups of three to practice active listening.

- Mentor #1 shares a current challenge that he or she is facing in a mentoring relationship.
- Mentor #2 practices active listening skills and tries to help mentor #1 develop a plan of action to resolve the situation.
- Mentor #3 acts as an observer and notes tone, body language, facial expression, and so on.
- Mentor #1 should share for 5 minutes with mentor #2. After 5 minutes, mentor #3 should provide feedback to both mentor #1 and mentor #2.

NOTE ▶ Encourage listeners to ask open-ended questions to explore the situation, and try to summarize or paraphrase what the speaker is saying to confirm understanding ("What I heard you say was . . .").

DISCUSS ▶ (10 min) With the entire group, have mentors share what they learned from the exercise and the strategies that the groups elicited. You may want to record the ideas generated in this discussion on the whiteboard or flip chart. Consider separating general comments from specific strategies for improving communication.

Postsession Assignment

TELL ▶ Have participants write a brief description of how they and their mentee are different. If they don't have a mentee, have them give a brief description of a past academic or work experience in which differences between individuals played a major role. Encourage mentors not to take the easy way out and talk about mere differences of opinion but to look deeper into relationships and backgrounds and how these differences might affect their mentoring relationship.

Mentoring Tool*

The following mentoring tool is located at the end of the chapter:

- Reflecting on Your Mentoring Relationship

This tool is intended to help students reaffirm their goals and expectations with their research mentor(s) and discuss any midproject changes that they would like to make.

* This tool is adapted from Branchaw, J. L., Pfund, C., and Rediske, R. (2010), *Entering Research: A Facilitator's Manual: Workshops for Students Beginning Research in Science*, W.H. Freeman & Company.

FACILITATION GUIDE

Case Study: The Slob

A graduate student mentor was frustrated because her undergraduate student mentee was not running successful experiments. While the undergraduate student had great enthusiasm for the project, each experiment failed because of some sloppy error: forgetting to pH the gel buffer, forgetting to add a reagent to a reaction, or forgetting to turn down the voltage on a gel box.

After a month of discussions, and careful attempts to teach the undergraduate student habits that would compensate for forgetfulness, the graduate student mentor was ready to give up. She spoke with her faculty adviser (the PI in the lab) and asked for advice, hoping that she could fix the problem. The adviser offered to work with the undergraduate student mentee. When the undergraduate student walked into his office the next day, the faculty adviser said, "I hear you're a slob in the lab. You gotta clean up your act if we're going to get any data out of you." Seeing the crushed and humiliated look on the student's face, he quickly added, "I'm a slob too—that's why I'm in here pushing papers around and not in the lab doing the hard stuff like you guys!"

Guiding Questions for Discussion:

1. If you were the mentee, how would you feel?

2. If you were the mentor, how would you feel?

3. If you were the faculty adviser, how would you feel?

4. If you were the adviser, how would you have handled this situation?

Mentoring Tool

Reflecting on Your Mentoring Relationship[*]

> **Objective:** Students will reaffirm their goals and expectations with their research mentor(s) and discuss any mid-project changes that they would like to make.

Maintaining a positive relationship with your research mentor is very important and can be achieved through frequent, open, and honest communication. To facilitate this communication, answer the questions below, then meet with your mentor to discuss them. You may also give a copy of the questions to your mentor to reflect on before the meeting.

1. What seems to be working well for you in the mentor-mentee relationship?

2. What is not working so well for you?

3. Review the goals and expectations you established with your mentor at the beginning of your relationship. Do you still agree that these goals and expectations are appropriate for your research experience, or do they need to be adjusted? Are you satisfied with the rate of progress you have made toward reaching the goals? If not, what might you do differently?

4. What has the relationship you have with your mentor taught you about what you must do to be successful as a researcher?

5. What aspects of mentoring do you need to get from someone other than your direct mentor? Who can provide this mentoring?

Write a paragraph summarizing the conversation you had with your mentor.

---

[*]Adapted from Branchaw, J. L., Pfund, C., and Rediske, R. (2010), *Entering Research: A Facilitator's Manual: Workshops for Students Beginning Research in Science*, W.H. Freeman & Company.

CHAPTER 5

Addressing Equity and Inclusion

Introduction

Diversity, in many dimensions, offers both challenges and opportunities to any relationship. Learning to identify, reflect upon, learn from, and engage with diverse perspectives is essential to fostering effective mentoring relationships and vibrant intellectual environments.

Learning Objectives

Mentors will have the knowledge and skills to

1. **Increase understanding of equity and inclusion and their influence on mentor-mentee interactions**
2. **Recognize the impact of conscious and unconscious assumptions, preconceptions, biases, and prejudices on the mentor-mentee relationship and acquire skills to manage them**
3. **Identify concrete strategies for learning about and addressing issues of equity and inclusion**

Overview of Activities for the "Addressing Equity and Inclusion" Session

	Learning Objectives	Core Activities
1	Increase understanding of equity and inclusion and their influence on mentor-mentee interactions	Consider the many ways mentors are and can be different from their mentees and how these differences affect the mentoring experience for both (Activity #1)
2	Recognize the impact of conscious and unconscious assumptions, preconceptions, biases, and prejudices on the mentor-mentee relationship and acquire skills to manage them	Reflect on unconscious assumptions (Activity #2) Read the results of diversity studies, discuss implications, and brainstorm strategies for reducing bias (Activity #3)
3	Identify concrete strategies for learning about and addressing issues of equity and inclusion	Read and discuss the case study, "Is It OK to Ask?" (Activity #4)

FACILITATION GUIDE

Recommended Session for Addressing Equity and Inclusion (60 minutes)

Materials Needed for the Session:

- Table tents
- Chalkboard, whiteboard, or flip chart
- Handouts:
  - Copies of introduction and learning objectives for Addressing Equity and Inclusion (page 63)
  - Copies of the "Diversity Study Results" handout (pages 68–69)
  - Copies of the case study, "Is It OK to Ask?" (page 70)
  - Copies of the reading, "Benefits and Challenges of Diversity" (pages 71–82)

Introductions (5 min)

TELL ▶ Review the introduction and learning objectives for the session.

Objective 1: Increase understanding of equity and inclusion and their influence on mentor-mentee interactions (10 min)

ACTIVITY #1: Reflecting on Diversity (10 min)

ASK ▶ Have the participants share one of the differences between them and their mentee that they identified for this week's assignment. Ask how these differences might impact their mentoring relationships.

TELL ▶ Acknowledge that diversity is often used in reference to race and ethnicity, but remember that it is broader than that. For example, consider the impact of learning and physical disabilities, gender, age/generation, professional experience, sexual orientation, class, and religion, and differences in communication, learning, and work styles.

Objective 2: Recognize the impact of conscious and unconscious assumptions, preconceptions, biases, and prejudices on the mentor-mentee relationship and acquire skills to manage them (28 min)

ACTIVITY #2: Reflect on Unconscious Assumptions (10 min)

TELL ▶ Open the activity by telling participants the following: Think about some of your assumptions when you entered the room on the first day of this training—that there would be electricity, a table, a bathroom, and so on. Let's think about some of the assumptions we make about the people we work with.

TELL ▶ (3 min) Read each word on the list below and ask mentors to focus on the first image that comes to their mind and to quickly jot down three words that describe the person they pictured. Pacing is important: leave only about 5 seconds between each item on the list so that participants are focused on the first image that comes to mind.
  * Cook
  * Mountain climber
  * Caretaker
  * Politician
  * Researcher
  * Graduate student
  * Undergraduate student
  * PI

DISCUSS ▶ (7 min) Have mentors share with the rest of the group some of the words they noted about each prompt, with special attention given to the undergraduate student. For example, did their images include mention of gender, race, body shape and size, or age? Was there some uniformity in their images?

TELL ▶ Remind mentors that we all carry these unconscious assumptions and they need not be a source of guilt or embarrassment. We discuss them as a means of raising awareness so that we can be conscious of them and minimize their impact on our behavior. The following studies highlight how enculturation affects us all and how it may impact the mentoring relationship.

ACTIVITY #3: Implications of Diversity Research (18 min)

READ ▶ (3 min) Distribute the "Diversity Study Results" handout (pages 68–69) and let participants read it individually.

NOTE ▶ Many of these studies are summarized in the reading "Benefits and Challenges of Diversity," which is included in the materials handed out.

DISCUSS ▶ (7 min) Have participants pair off to discuss their reactions to one of the studies and the implications for their mentoring practice.

NOTE ▶ You may ask mentors to choose a study to discuss, or you may wish to assign one study to each pair.

DISCUSS ▶ (8 min) Engage the entire group in discussion. You may want to record the ideas generated in this discussion on the whiteboard or flip chart. Guide the discussion using the following questions:

1. What were your initial reactions to the studies?
2. Which study captured your attention? Why?
3. What implications do these study results have for your mentoring practice?
4. What are two to three practical things you could do to minimize the impact of bias, prejudice, and stereotype in your mentoring relationship?

Objective 3: Identify concrete strategies for learning about and addressing issues of equity and inclusion (17 min)

ACTIVITY #4: Case Study: Is It OK to Ask?

READ ▶ (2 min) Distribute the case study and ask a participant to read the case aloud.

DISCUSS ▶ (15 min) Discuss the case study with the entire group. You may want to record the ideas generated in this discussion on the whiteboard or flip chart. Use the guiding questions following the case study to facilitate discussion. Here are some additional questions:

1. As a mentor, would you feel comfortable asking a mentee about how their racial or any other identity affects their experiences? How do you decide when asking questions about these issues is appropriate?
2. Specifically, how would you go about engaging someone in a discussion about their race, ethnicity, class, gender, disability, age, sexual orientation, background, or personal values? How do you engage in such conversations based on interest without feeling or expressing a sense of judgment about differences? What are the risks of addressing or not addressing diversity directly?
3. How can you view diversity as an asset to a mentor-mentee relationship? Reframe conversations with a mentee in terms of how you can benefit and learn from their experiences that differ from your own.
4. Do you think everyone should be treated the same? Does treating everyone the same mean they are being treated equally?

NOTE ▶ In some groups, mentors can be fairly quiet and reluctant to speak at first in this discussion, but just give them a few minutes. Once mentors get going with the discussion, it is often rich and engaging. Allowing mentors to choose which case they would like to discuss should help. Be sensitive to the fact that some minority mentors may be assumed to speak for all members of their group or all minorities, and this is offensive to some. Possible responses to the cases are included below.

NOTE ▶ Views of the impact of race, class, ethnicity, gender, disability, age, sexual orientation, and background on the research experience vary widely. Remember that as a facilitator *you are not expected to be an expert* on the topic. Given that some facilitators have expressed less comfort managing this session, we have compiled some comments that you may hear in response to the cases. Given the complexity of human relationships and the importance of situational contexts, these responses are of course by no means exhaustive or comprehensive, but they may prepare you to handle the range of views that will emerge. Possible responses to the case studies include the following:

- "Race, gender, class, ethnicity, sexual orientation, personal values, and other aspects of diversity have nothing to do with a research experience because the experience should focus on research and not on personal characteristics. It would not be fair to treat one lab member differently than another. The approach to the research must be objective and influenced as little as possible by the individual doing the research."

- "Race, gender, class, ethnicity, sexual orientation, personal values, and other aspects of diversity have everything to do with a research experience and permeate every aspect of the experience, impacting perceptions, confidence, and motivation. Ignoring the impact of diversity sends a message that those aspects of a person have no role in one's work, which may turn students off to science. The level of impact will vary across the relationship. At times it may be invisible. At other times, it may be the most important factor."

- "Individuals want to be assessed for their ability, independent of race, gender, and other aspects of diversity. The trick is deciding how to balance acknowledging someone's background and taking it into consideration when deciding how to work with that person, but not letting a person's background bias your interaction with them."

- "Regular conversations with *all* mentees to check on how they are doing and whether they are happy in their overall environment are important. This builds relationships that allow mentees to be comfortable sharing concerns; it also allows mentors to notice if there are issues surrounding race or other diverse personal characteristics that need to be addressed, or identify opportunities for growth."

Postsession Assignments

TELL ▶ Advise participants to reflect on their compacts and IDPs and identify areas where they should consider revising them based on their reflections on equity and inclusion.

TELL ▶ Have everyone read "Benefits and Challenges of Diversity."

FACILITATION GUIDE

Diversity Study Results*

Read the description of the study results and discuss your reaction and the implications for your mentoring practice. See the reading "Benefits and Challenges of Diversity," included in this session, for more details about these and other studies.

Study 1: Studies of hiring involve assigning a man's name or woman's name to the same application and randomly distributing the applications to a group of reviewers. The reviewers are more likely to hire the person if there is a man's name on the application. The sex of the reviewer has no effect on the outcome. The result has not changed much over 40 years of doing the study (Steinpreis, Anders, et al. 1999; Dovidio and Gaertner 2000; Moss-Racusin, Dovidio, et al. 2013).

Study 2: Many studies show that when reviewers are asked to review job performance based on a written description of the person's accomplishments, they rate the performance higher if they are told that they are reviewing a man. In one study the difference between ratings for men and women candidates was greater when the evaluator was busy or distracted. The sex of the reviewer was not significant (Martell and Leavitt 2002).

Study 3: A linguistic analysis of 300 letters of recommendation for successful candidates applying for (and ultimately being offered) faculty positions at a major medical school showed differences in language and content. Male candidates were referred to more often as "researchers" and "colleagues," whereas women were referred to as "teachers" and "students." There were four times more references to women's personal lives than to men's, and there were more "doubt raisers" in letters about women (Trix and Psenka 2003).

Study 4: An ecology journal initiated a double-blind review (authors' names not revealed to reviewers, reviewers' names not revealed to authors). During the 6-month period of the trial, the acceptance rate for papers first-authored by women increased significantly. There was no change in the frequency of acceptance of papers first-authored by women in a similar ecology journal during the same period (Budden, Tregenza, et al. 2008).

Study 5: Evaluators expressed less prejudice against African American candidates if they were instructed to avoid prejudice (Lowery, Hardin, et al. 2001).

Study 6: When participants were shown images of admired black figures, they associated negative words with black people less than those who were shown pictures of disliked black figures or not shown pictures at all (Blair, Ma, et al. 2001; Dasgupta and Greenwald 2001).

Study 7: Subjects were told to select one of two rooms in which to watch a movie. In each situation there is a handicapped person sitting in one of the rooms. If both rooms are showing the same movie, the subjects were more likely to choose the room where the handicapped person is sitting. If the

---

*Many of these studies and others are summarized in Fine and Handelsman, "Benefits and Challenges of Diversity," which appears at the end of this chapter, and Handelsman, Miller, and Pfund (2005), "Diversity," in *Scientific Teaching*, New York: W.H. Freeman & Company. This activity was taken from the National Academies Summer Institute on Undergraduate Education in Biology (http://www.academiessummerinstitute.org, accessed June 2010).

rooms are showing different movies, the subjects are more likely to choose the room where the handicapped person is not sitting. The result is the same independent of which movie is showing in the room with the handicapped person (Snyder 1979).

Study 8: One study examined differences over a 10-year period of whites' self-reported racial prejudice and their bias in selection decisions involving black and white candidates for employment. They report that self-reported prejudice was lower in 1998–1999 than it was in 1988–1989. At both time points, white participants did not discriminate against black candidates when their qualifications were clearly strong or weak, but they did discriminate when the qualifications were mixed or the decision ambiguous (Dovidio and Gaertner 2000).

Study 9: Stereotype threat is the anxiety people feel about confirming stereotypes of a group to which they belong. When stereotype threat is activated, usually by reminding a person of their race or sex, a person may identify with a negative stereotype and perform less well than without activation. MRI examination of the human brain shows that activating stereotype threat makes blood move from the cognitive centers to the affective centers of the brain (Krendl, Richeson, et al. 2008).

Study 10: A wide range of studies show that racial and ethnic minorities tend to receive lower-quality health care and are less likely to receive routine medical procedures than nonminority patients, even when the issue of access to health care is controlled (Smedley, Stith, and Nelson 2003).

Study References

Blair, I. V., J. E. Ma, et al. (2001). "Imagining stereotypes away: The moderation of implicit stereotypes through mental imagery." *J Pers Soc Psychol* 81(5): 828–841.

Budden, A. E., T. Tregenza, et al. (2008). "Double-blind review favours increased representation of female authors." *Trends in Ecology & Evolution* (Personal edition) 23(1): 4–6.

Dasgupta, N., and A. G. Greenwald (2001). "On the malleability of automatic attitudes: Combating automatic prejudice with images of admired and disliked individuals." *J Pers Soc Psychol* 81(5): 800–814.

Dovidio, J. F., and S. L. Gaertner (2000). "Aversive racism and selection decisions: 1989 and 1999." *Psychological Science* 11: 319.

Krendl, A. C., J. A. Richeson, et al. (2008). "The negative consequences of threat—A functional magnetic resonance imaging investigation of the neural mechanisms underlying women's underperformance in math." *Psychological Science* 19(2): 168–175.

Lowery, B. S., C. D. Hardin, et al. (2001). "Social influence effects on automatic racial prejudice." *J Pers Soc Psychol* 81(5): 842–855.

Martell, R. F., and K. N. Leavitt (2002). "Reducing the performance-cue bias in work behavior ratings: Can groups help?" *J Appl Psychol* 87(6): 1032–1041.

Moss-Racusin, C. A., J. F. Dovidio, et al. (2013). "Science faculty's subtle gender biases favor male students." *Proceedings of the National Academy of Sciences* 109(41): 16474–16479.

Smedley, B. D., A. Y. Stith, and A. R. Nelson. (2003). *Unequal Treatment: Confronting Racial and Ethnic Disparities*. Washington, DC: National Academies Press.

Snyder, M. L. (1979). "Avoidance of the handicapped—Attibutional ambiguity analysis." *J Pers Soc Psychol* 37(12): 2297–2306.

Steinpreis, R. E., K. A. Anders, et al. (1999). "The impact of gender on the review of the curricula vitae of job applicants and tenure candidates: A national empirical study." *Sex Roles* 41(7/8): 509–528.

Trix, F., and C. Psenka (2003). "Exploring the color of glass: Letters of recommendation for female and male medical faculty." *Discourse & Society* 14(2): 191–220.

Case Study: Is It OK to Ask?

Last summer I worked with a fantastic undergraduate mentee. She was very intelligent and generated a fair amount of data directly relevant to my thesis project. I think that she had a positive summer research experience, but there are a few questions that still linger in my mind. This particular mentee was an African American woman from a small town. I always wondered how she felt on a big urban campus. I also wondered how she felt about being the only African American woman in our lab. In fact, she was the only African American woman in our entire department that summer. I wanted to ask her how she felt, but I worried that it might be insensitive or politically incorrect to do so. I never asked. I still wonder how she felt and how those feelings may have affected her experience.

Guiding Questions for Discussion:

1. What are the main themes raised in this case study?

2. What might the mentor's intent have been in asking the question, and what might the impact be on the mentee?

3. How might you react to this case differently if the mentee was the only openly gay student in the department? How do you engage in such conversations based on interest without feeling or expressing a sense of judgment about differences? How do you ask without raising issues of tokenism?

Benefits and Challenges of Diversity[*]

by Eve Fine and Jo Handelsman

The diversity of a university's faculty, staff, and students influences its strength, productivity, and intellectual personality. Diversity of experience, age, physical ability, religion, race, ethnicity, gender, and many other attributes contributes to the richness of the environment for teaching and research. We also need diversity in discipline, intellectual outlook, cognitive style, and personality to offer students the breadth of ideas that constitute a dynamic intellectual community.

A vast and growing body of research provides evidence that a diverse student body, faculty, and staff benefits our joint missions of teaching and research by increasing creativity, innovation, and problem solving. Yet diversity of faculty, staff, and students also brings challenges. Increasing diversity can lead to less cohesiveness, less effective communication, increased anxiety, and greater discomfort for many members of a community.[1] Learning to respect and appreciate each other's cultural and stylistic differences and becoming aware of unconscious assumptions and behaviors that may influence our interactions will enable us to minimize the challenges and derive maximum benefits from diversity.

This article summarizes research on the benefits and challenges of diversity and provides suggestions for realizing the benefits. Its goal is to help create a climate in which all individuals feel *personally safe, listened to, valued, and treated fairly and with respect.*[2]

> It is time to renew the promise of American higher education in advancing social progress, end America's discomfort with race and social difference, and deal directly with many of the issues of inequality present in everyday life.　　　　　　　　　　　　　　　　　　　　—Sylvia Hurtado

Benefits for Teaching and Research

Research shows that diverse working groups are more productive, creative, and innovative than homogeneous groups, and suggests that developing a diverse faculty will enhance teaching and research.[3] Here are some of the findings.

- A controlled experimental study of performance during a brainstorming session compared ideas generated by ethnically diverse groups composed of Asians, blacks, whites, and Latinos to those generated by ethnically homogenous groups composed of whites only. Evaluators who were unaware of the source of the ideas found no significant difference in the number of ideas generated by the two types of groups. However, when applying measures of feasibility and effectiveness, they rated the ideas generated by diverse groups as being of higher quality.[4]
- The level of critical analysis of decisions and alternatives was higher in groups exposed to minority viewpoints than in groups that were not. Minority viewpoints stimulated discussion of multiple perspectives and previously unconsidered alternatives, whether or not the minority opinion was correct or ultimately prevailed.[5]
- A study of corporate innovation found that the most innovative companies deliberately established diverse work teams.[6]
- Data from the 1995 Faculty Survey conducted by UCLA's Higher Education Research Institute (HERI) demonstrated that scholars from minority groups have expanded and enriched scholar-

[*] From Handelsman, J., Pfund, C., Miller Lauffer, S., and Pribbenow, C. M. (2005), *Entering Mentoring: A Seminar to Train a New Generation of Scientists,* Madison, WI: University of Wisconsin Press

READING

ship and teaching in many academic disciplines by offering new perspectives and by raising new questions, challenges, and concerns.[7]

- Several investigators found that women and faculty of color more frequently employed active learning in the classroom, encouraged student input, and included perspectives of women and minorities in their coursework.[8]

Benefits for Students

Numerous research studies have examined the impact of diversity on students and educational outcomes. Cumulatively, these studies provide extensive evidence that diversity has a positive impact on all students, minority and majority.[9] Here are some examples.

- A national longitudinal study of 25,000 undergraduates at 217 four-year colleges and universities showed that institutional policies fostering diversity of the campus community had positive effects on students' cognitive development, satisfaction with the college experience, and leadership abilities. These policies encouraged faculty to include themes relating to diversity in their research and teaching, and provided students with opportunities to confront racial and multicultural issues in the classroom and in extracurricular settings.[10]

- Two longitudinal studies, one conducted by HERI in 1985 and 1989 with over 11,000 students from 184 institutions and another in 1990 and 1994 with approximately 1,500 students at the University of Michigan, showed that students who interacted with racially and ethnically diverse peers both informally and within the classroom showed the greatest "engagement in active thinking, growth in intellectual engagement and motivation, and growth in intellectual and academic skills."[11] A more recent study of 9,000 students at 10 selective colleges reported that meaningful engagement rather than casual and superficial interactions led to greater benefit from interaction with racially diverse peers.[12]

- Data from the National Study of Student Learning indicated that both in-class and out-of-class interactions and involvement with diverse peers fostered critical thinking. This study also found a strong correlation between "the extent to which an institution's environment is perceived as racially nondiscriminatory" and students' willingness to accept both diversity and intellectual challenge.[13]

- A survey of 1,215 faculty members in departments granting doctoral degrees in computer science, chemistry, electrical engineering, microbiology, and physics showed that women faculty played important roles in fostering the education and success of women graduate students.[14]

Challenges of Diversity

Despite the benefits that a diverse faculty, staff, and student body provide to a campus, diversity also presents considerable challenges that must be addressed and overcome. Here are some examples.

- Numerous studies have reported that women and minority faculty members are considerably less satisfied with many aspects of their jobs than are majority male faculty members. These aspects include teaching and committee assignments, involvement in decision making, professional relations with colleagues, promotion and tenure, salary inequities, and overall job satisfaction.[15]

- A study of minority faculty at universities and colleges in eight midwestern states showed that faculty of color experience exclusion, isolation, alienation, and racism in predominantly white universities.[16]

- Multiple studies demonstrate that minority students often feel isolated and unwelcome in predominantly white institutions and that many experience discrimination and differential treatment. Minority status can result from race, ethnicity, national origin, sexual orientation, disability, and other factors.[17]

- Women students, particularly when they are minorities in their classes, may experience unwelcoming climates that can include sexist use of language, presentation of stereotypic or disparaging views of women, differential treatment from professors, and/or sexual harassment.[18]

- When a negative stereotype relevant to their identity exists in a field of interest, women and members of minority groups often experience "stereotype threat"—the fear that they will confirm or be judged in accordance with the stereotype. Such stereotype threat exists both for entry into a new field and for individuals already excelling in a specific arena. Situations or behaviors that heighten awareness of one's minority status can activate stereotype threat.[19] Research demonstrates that once activated, stereotype threat leads to stress and anxiety, which decreases memory capacity, impairs performance, and reduces aspirations and motivation.[20] Human brain imaging, which shows that activating stereotype threat causes blood to move from the cognitive to the affective centers of the brain, indicates how situational cues reduce cognitive abilities.[21]

- Research has demonstrated that a lack of previous positive experiences with "outgroup members" (minorities) causes "ingroup members" (majority members) to feel anxious about interactions with minorities. This anxiety can cause majority members to respond with hostility or to avoid interactions with minorities.[22]

Influence of Unconscious Assumptions and Biases

Research studies show that people who have strong egalitarian values and believe that they are not biased may unconsciously behave in discriminatory ways.[23] A first step toward improving climate is to recognize that unconscious biases, attitudes, and other influences unrelated to the qualifications, contributions, behaviors, and personalities of our colleagues can influence our interactions, *even if we are committed to egalitarian views*. Although we all like to think that we are objective scholars who judge people on merit, the quality of their work, and the nature of their achievements, copious research shows that a lifetime of experience and cultural history shapes every one of us and our judgments of others.

> People confident in their own objectivity may overestimate their invulnerability to bias.
>
> —Eric Uhlmann and Geoffrey Cohen

The results from controlled research studies demonstrate that people often hold unconscious, implicit assumptions that influence their judgments and interactions with others. Examples range from expectations or assumptions about physical or social characteristics associated with race, gender, age, and ethnicity to those associated with certain job descriptions, academic institutions, and fields of study. Let's start with some examples of common social assumptions or expectations.

- When shown photographs of people of the same height, evaluators overestimated the heights of male subjects and underestimated the heights of female subjects, even though a reference point, such as a doorway, was provided.[24]

- When shown photographs of men of similar height and build, evaluators rated the athletic ability of black men higher than that of white men.[25]

- When asked to choose counselors from a group of equally competent applicants who were neither exceptionally qualified nor unqualified for the position, college students chose white candidates more often than African American candidates, exhibiting a tendency to give members of the majority group the benefit of the doubt.[26]

READING

These studies show that we often apply generalizations about groups that may or may not be valid to the evaluation of individuals.[27] In the study on height, evaluators applied the statistically accurate generalization that men are usually taller than women to estimate the height of individuals who did not necessarily conform to the generalization. If we can inaccurately apply generalizations to objective characteristics as easily measured as height, what happens when the qualities we are evaluating are not as objective or as easily measured? What happens when, as in the studies of athletic ability and choice of counselor, the generalizations are not valid? What happens when such generalizations unconsciously influence the ways we interact with other people? Here are some examples of assumptions or biases that can influence interactions.

- When rating the quality of verbal skills as indicated by vocabulary definitions, evaluators rated the skills lower if told that an African American provided the definitions than if told that a white person provided them.[28]
- When asked to assess the contribution of skill versus luck to successful performance of a task, evaluators more frequently attributed success to skill for males and to luck for females, even though males and females performed the task identically.[29]
- Evaluators who were busy, distracted by other tasks, and under time pressure gave women lower ratings than men for the same written evaluation of job performance. Sex bias decreased when they took their time and focused attention on their judgments, which rarely occurs in actual work settings.[30]
- Research has shown that incongruities between perceptions of female gender roles and leadership roles can cause evaluators to assume that women will be less competent leaders. When women leaders provided clear evidence of their competence, thus violating traditional gender norms, evaluators perceived them to be less likable and were less likely to recommend them for hiring or promotion.[31]
- A study of nonverbal communication found that white interviewers maintained higher levels of visual contact, reflecting greater attraction, intimacy, and respect, when talking with white interviewees and higher rates of blinking, indicating greater negative arousal and tension, when talking with black interviewees.[32]

Several research studies conclude that implicit biases and assumptions can affect evaluation and hiring of candidates for academic positions. These studies show that the gender of the person being evaluated significantly influences the assessment of résumés and postdoctoral applications, evaluation of journal articles, and the language and structure of letters of recommendation. As we attempt to enhance campus and department climate, the influence of such biases and assumptions may also affect selection of invited speakers and conference presenters, committee membership, interaction and collaboration with colleagues, and promotion to tenure and full professorships. Here are some examples of assumptions or biases in academic contexts.

- A study of over 300 recommendation letters for medical faculty hired by a large American medical school found that letters for female applicants differed systematically from those for males. Letters written for women were shorter, provided "minimal assurance" rather than solid recommendations, raised more doubts, and included fewer superlative adjectives.[33]
- In a national study, 238 academic psychologists (118 male, 120 female) evaluated a junior-level or a senior-level curriculum vitae randomly assigned a male or a female name. These were actual vitae from an academic psychologist who successfully competed for an assistant professorship and then received tenure early. For the junior-level applicant, both male and female evaluators gave

the male applicant better ratings for teaching, research, and service and were more likely to hire the male than the female applicant. Gender did not influence evaluators' decisions to tenure the senior-level applicant, but evaluators did voice more doubts about the female applicant's qualifications.[34]

- A study of postdoctoral fellowships awarded by the Medical Research Council of Sweden found that women candidates needed substantially more publications to achieve the same rating as men, unless they personally knew someone on the selection panel.[35]

- A 2008 study showed that when the journal *Behavioral Ecology* introduced a double-blind review process that concealed the identities of reviewers and authors, there was a significant increase in the publication of articles with a woman as the first author.[36]

Reaping the Benefits and Minimizing the Challenges of Diversity

To reap the benefits and minimize the challenges of diversity, we need to overcome the powerful human tendency to feel more comfortable when surrounded by people we resemble. We need to learn how to understand, value, and appreciate difference. Here is some advice for doing so.

Become aware of unconscious biases that may undermine your conscious commitment to egalitarian principles.

One way of doing this is to take the Implicit Association Test (IAT) offered by Project Implicit, a research collaborative at the University of Virginia, Harvard University, and the University of Washington (https://implicit.harvard.edu/implicit/demo).

Consciously strive to minimize the influence of unintentional bias.

Question your judgments and decisions and consider whether unintentional bias may have played a role. One way to do so is to perform a thought experiment: ask yourself if your opinions or conclusions would change if the person were of a different race, sex, religion, and so forth. Some questions to consider include the following:

- Are women or minority colleagues/students subject to higher expectations in areas such as number and quality of publications, name recognition, or personal acquaintance with influential colleagues?

- Are colleagues or students who received degrees from institutions other than major research universities undervalued? Are we missing opportunities to benefit from the innovative, diverse, and valuable perspectives and expertise of colleagues or students from other institutions such as historically black universities, four-year colleges, community colleges, government, or industry?

- Are ideas and opinions voiced by women or minorities ignored? Are their achievements and contributions undervalued or unfairly attributed to collaborators, despite evidence to the contrary in their publications or letters of reference?

- Is the ability of women or minorities to lead groups, raise funds, and/or supervise students and staff underestimated? Are such assumptions influencing committee and/or course assignments?

- Are assumptions about whether women or minorities will "fit in" to an existing environment influencing decisions?

- Are assumptions about family obligations inappropriately influencing appointments and other decisions?

READING

Seek out opportunities for greater interaction with women and minority colleagues.
Get to know women and minority colleagues in your department, your campus, and your professional associations. Pursue meaningful discussions with them about research, teaching methodologies, and ideas about the direction of your department, college, and profession. Listen actively to any concerns they express and try to understand and learn from their perspectives and experiences.

Focus on the individual and on their personality, qualifications, merit, and interests.
Consciously avoid the tendency to make assumptions about an individual based on the characteristics (accurate or not) of their group membership. Likewise, avoid the tendency to make assumptions about groups based on the behavior, personality, or qualifications of an individual group member. Instead, concentrate on the individual and their qualities.

Treat all individuals—regardless of race, sex, or status—with respect, consideration, and politeness.
- Greet faculty, staff, and students pleasantly in hallways or in other chance encounters.
- Make requests to faculty, staff, and students politely—even when the work you are asking for is part of their obligations.
- Acknowledge and appreciate the work, assistance, and contributions of faculty colleagues, staff, and students. Do so in public forums as well as privately.
- Address individuals by their appropriate titles or by their preferred forms of address.

Actively promote inclusive communities.
- In classroom, committee, laboratory, and departmental settings, work to ensure that everyone has a chance to voice opinions, concerns, or questions. Acknowledge and attribute ideas, suggestions, and comments accurately. Women and minorities often report that their remarks or contributions are ignored or unheard.
- Support efforts to ensure that leadership and membership of departmental and professional committees are diverse with respect to age, gender, nationality, race, ethnicity, and so on.
- Support efforts to ensure that departmental events such as seminar series and sponsored conferences include presenters of various ages, genders, nationalities, races, and ethnicities.
- Promote inclusive language by example. Avoid using only male pronouns when referring to groups of both sexes. Avoid language that makes assumptions about marital status and or/sexual orientation; for example, consider using "partner" rather than "spouse."
- Welcome new departmental members by initiating conversations or meetings with them. Attend social events hosted by your department and make efforts to interact with new members and others who are not part of your usual social circle.

Avoid activating stereotype threat.
In addition to the preceding advice for actively promoting inclusive communities, the following suggestions can prevent the activation of stereotype threat or counteract its effects:

- Teach students and colleagues about stereotype threat.[37]
- Counter common stereotypes by increasing the visibility of successful women and minority members of your discipline. Ensure that the posters and/or photographs of members of your department or discipline displayed in hallways, conference rooms, and classrooms reflect the diversity you wish to achieve. Choose textbooks that include the contributions and images of diverse members of your discipline.[38]

- Support and encourage your students by providing positive feedback as well as constructive criticism to ensure that they know their strengths and develop confidence in their abilities. Save your harshest criticism for private settings so that you do not humiliate or embarrass students in front of either their peers or more senior colleagues. Such respectful practices are important for all students, but are likely to be more important for women and members of minority groups, who may have received less encouragement and may be at greater risk of being discouraged due to the influence of stereotype threat. Demonstrate similar respect and encouragement for your colleagues.
- For more suggestions, see http://reducingstereotypethreat.org/reduce.html.

Conclusion

Diversity is not an end in itself. Diversity is a means of achieving our educational and institutional goals. As such, merely adding diverse people to a homogeneous environment does not automatically create a more welcoming and intellectually stimulating campus.

Long-term efforts, engagement, and substantial attention are essential for realizing the benefits that diversity has to offer and for ensuring that all members of the academic community are respected, listened to, and valued.

References

1. Manzoni, J.-F., P. Strebel, and J.-L. Barsoux. "Why diversity can backfire on company boards." *MIT Sloan Management Review—Business Insight*, January 22, 2010.

   Herring, C. "Does diversity pay? Race, gender, and the business case for diversity." *American Sociological Review* 74 (2009): 208–224.

   Page, S. E. (2007). *The Difference: How the Power of Diversity Creates Better Groups, Firms, Schools, and Societies.* Princeton, NJ: Princeton University Press.

   Putnam, R. D. "*E Pluribus Unum*: Diversity and community in the twenty-first century—The 2006 Johan Skytte Prize Lecture." *Scandinavian Political Studies* 30 (2007): 137–174.

   van Knippenberg, D., and M. C. Schippers. "Work group diversity." *Annual Review of Psychology* 58 (2007): 515–541.

   Mannix, E., and M. A. Neale. "What differences make a difference? The promise and reality of diverse teams in organizations." *Psychological Science in the Public Interest* 6 (2005): 31–55.

   Cox, T., Jr. (1993). *Cultural Diversity in Organizations: Theory, Research & Practice.* San Francisco: Berrett-Koehler Publishers.

2. University of Wisconsin–Madison, Office of the Provost (2004). "Definition of Campus Climate." (http://wiseli.engr.wisc.edu/climate/Provost_ClimateDefn.pdf)

3. Herring, C. "Does diversity pay? Race, gender, and the business case for diversity." *American Sociological Review* 74 (2009): 208–224.

   Chang, M. J., D. Witt, et al. (2003). *Compelling Interest: Examining the Evidence on Racial Dynamics in Colleges and Universities.* Stanford, CA: Stanford University Press.

   American Council on Education (ACE) and American Association of University Professors (AAUP) (2000). *Does Diversity Make a Difference? Three Research Studies on Diversity in College Classrooms.* Washington, DC: ACE and AAUP.

4. McLeod, P. L., S. A. Lobel, and T. H. Cox Jr. "Ethnic diversity and creativity in small groups." *Small Group Research* 27 (1996): 248–265.

READING

5. Nemeth, C. J. "Dissent as driving cognition, attitudes, and judgments." *Social Cognition* 13 (1995): 273–291.

   Nemeth, C. J. "Differential contributions of majority and minority influence." *Psychological Review* 93 (1986): 23–32.

   Nemeth, C. J. "Dissent, group process, and creativity: The contribution of minority influence." *Advances in Group Process* 2 (1985): 57–74.

   Schulz-Hardt, S. et al. "Group decision making in hidden profile situations: Dissent as a facilitator for decision quality." *Journal of Personality and Social Psychology* 91 (2006): 1080–1093.

   Sommers, S. R. "On racial diversity and group decision making: Identifying multiple effects of racial composition on jury deliberations." *Journal of Personality and Social Psychology* 90 (2006): 597–612.

   Antonio, A. L., et al. "Effects of racial diversity on complex thinking in college students." *Psychological Science* 15 (2004): 507–510.

6. Kanter, R. M. (1983). *The Change Masters: Innovations for Productivity in the American Corporation*. New York: Simon and Schuster.

7. Antonio, A. L. "Faculty of color reconsidered: Reassessing contributions to scholarship." *Journal of Higher Education* 73 (2002): 582–602.

   Turner, C. S. V. "New faces, new knowledge." *Academe* 86 (2000): 34–37.

   Nelson, S., and G. Pellet (1997). *Shattering the Silences* [videorecording]. San Francisco: Gail Pellet Productions.

8. Milem, J. F. (2003). "The educational benefits of diversity: Evidence from multiple sectors." In *Compelling Interest: Examining the Evidence on Racial Dynamics in Colleges and Universities*, edited by M. J. Chang, D. Witt, J. Jones, and K. Hakuta, 126–169. Stanford, CA: Stanford University Press.

9. Smith, D. G., et al. (1997). *Diversity Works: The Emerging Picture of How Students Benefit*. Washington, DC: Association of American Colleges and Universities.

10. Astin, A. W. "Diversity and multiculturalism on the campus: How are students affected?" *Change* 25(2) (1993): 44–49.

    Astin, A. W. (1993). *What Matters in College? Four Critical Years Revisited*. San Francisco: Jossey-Bass.

11. Gurin, P., E. L. Dey, et al. "Diversity and higher education: Theory and impact on educational outcomes." *Harvard Educational Review* 72 (2002): 330–366.

    Gurin, P. "Selections from *The Compelling Need for Diversity in Higher Education*, expert reports in defense of the University of Michigan." *Equity & Excellence in Education* 32 (1999): 36–62.

12. Espenshade, T. J., and A. W. Radford (2009). *No Longer Separate, Not Yet Equal: Race and Class in Elite College Admission and Campus Life*. Princeton, NJ: Princeton University Press.

13. Pascarella, E. T., et al. "Influences on students' openness to diversity and challenge in the first year of college." *Journal of Higher Education* 67 (1996): 174–195.

14. Fox, M. F. (2003). "Gender, faculty, and doctoral education in science and engineering." In *Equal Rites, Unequal Outcomes: Women in American Research Universities*, edited by L. S. Hornig, 91–109. New York: Kluwer Academic/Plenum Publishers.

    Carbonell, J. L., and Y. Castro. "The impact of a leader model on high dominant women's self-selection for leadership." *Sex Roles* 58 (2008): 776–783.

    Kutob, R. M., J. H. Senf, and D. Campos-Outcalt. "The diverse functions of role models across primary care specialties." *Family Medicine* 38 (2006): 244–251.

    Bakken, L. L. "Who are physician-scientists' role models? Gender makes a difference." *Academic Medicine* 80 (2005): 502–506.

READING

15. Sheridan, J., and J. Winchell (2006). *Results from the 2006 Study of Faculty Worklife at UW–Madison.* Madison, WI: WISELI.

Sheridan, J., and J. Winchell (2003). *Results from the 2003 Study of Faculty Worklife at UW–Madison,* Madison, WI: WISELI.

Harvard University Task Force on Women Faculty (2005). *Report of the Task Force on Women Faculty.* Cambridge, MA: Harvard University.

Astin, H. S., and C. M. Cress (2003). "A national profile of academic women in research universities." In *Equal Rites, Unequal Outcome: Women in American Research Universities,* edited by L. S. Hornig, 53–88. New York: Kluwer Academic/Plenum Publishers.

Zakian, V., et al. (2003). *Report of the Task Force on the Status of Women Faculty in the Natural Sciences and Engineering at Princeton.* Princeton, NJ: Princeton University Press.

Allen, W. R., et al. (2002)."Outsiders within: Race, gender, and faculty status in US higher education." In *The Racial Crisis in American Higher Education: Continuing Challenges for the Twenty-First Century,* edited by W. A. Smith, P. G. Altbach, and K. Lomotey, 189–220. Albany, NY: State University of New York Press.

Trower, C. A., and R. P. Chait. "Faculty diversity: Too little for too long." *Harvard Magazine* 104 (2002): 33–37, 98.

Turner, C. S. V. "Women of color in academe: Living with multiple marginality." *Journal of Higher Education* 73 (2002): 74–93.

Aguirre, A., Jr. "Women and minority faculty in the academic workplace: Recruitment, retention, and academic culture." *ASHE-ERIC Higher Education Reports* 27 (2000).

Foster, S. W., et al. "Results of a gender-climate and work-environment survey at a midwestern academic health center." *Academic Medicine* 75 (2000): 653–660.

Turner, C. S. V., and S. L. Myers Jr. (2000). *Faculty of Color in Academe: Bittersweet Success.* Boston, MA: Allyn & Bacon.

MIT Committee on Women Faculty (1999). *A Study on the Status of Women Faculty in Science at MIT.* Boston, MA: Massachusetts Institute of Technology.

Blackburn, R. P., and C. Hollenshead (1999). *University of Michigan Faculty Work-Life Study Report.* Ann Arbor, MI: University of Michigan.

Riger, S., J. Stokes, et al. "Measuring perceptions of the work environment for female faculty." *Review of Higher Education* 21 (1997): 63–78.

16. Turner, C. S. V., and S. L. Myers Jr. (2000). *Faculty of Color in Academe: Bittersweet Success.* Boston, MA: Allyn & Bacon.

Turner, C. S. V. "Women of color in academe: Living with multiple marginality." *Journal of Higher Education* 73 (2002): 74–93.

17. Rankin, S. R. (2003). *Campus Climate for Gay, Lesbian, Bisexual, and Transgender People: A National Perspective.* New York: National Gay and Lesbian Task Force Policy Institute.

Suarez-Balcazar, Y., et al. "Experiences of differential treatment among college students of color." *Journal of Higher Education* 74 (2003): 428–444.

Hurtado, S., D. F. Carter, and D. Kardia. "The climate for diversity: Key issues for institutional self-study." *New Directions for Institutional Research* 98 (1998): 53–63.

Cress, C. M., and L. J. Sax. "Campus Climate Issues to Consider for the Next Decade." *New Directions for Institutional Research* 98 (1998): 65–80.

READING

Nora, A., and A. F. Cabrera. "The role of perceptions of prejudice and discrimination on the adjustment of minority students to college." *Journal of Higher Education* 67 (1996): 119–148.

Smedley, B. D., H. F. Myers, and S. P. Harrell. "Minority-status stresses and the college adjustment of ethnic minority freshmen." *Journal of Higher Education* 64 (1993): 434–452.

Hurtado, S. "The campus racial climate: Contexts of conflict." *Journal of Higher Education* 63 (1992): 539–569.

18. Salter, D. W., and A. Persaud. "Women's views of the factors that encourage and discourage classroom participation." *Journal of College Student Development* 44 (2003): 831–844.

Crombie, G., et al. "Students' perceptions of their classroom participation and instructor as a function of gender and context." *Journal of Higher Education* 74 (2003): 51–76.

Swim, J. K., L. L. Hyers, et al. "Everyday sexism: Evidence for its incidence, nature, and psychological impact from three daily diary studies." *Journal of Social Issues* 57 (2001): 31–53.

Whitt, E. J., et al. "Women's perceptions of a 'chilly climate' and cognitive outcomes in college: Additional evidence." *Journal of College Student Development* 40 (1999): 163–177.

Sands, R. G. "Gender and the perception of diversity and intimidation among university students." *Sex Roles* 39 (1998): 801–815.

Foster, T. J., et al. "An empirical test of Hall and Sandler's 1982 report: Who finds the classroom climate chilly?" Paper presented at the annual meeting of the Central States Communication Association, Oklahoma City, OK, April 1994.

Hall, R. M., and B. R. Sandler (1982). *The Classroom Climate: A Chilly One for Women?* Washington, DC: Project on the Status and Education of Women, Association of American Colleges.

19. Spencer, S. J., C. M. Steele, and D. M. Quinn. "Stereotype threat and women's math performance." *Journal of Experimental Social Psychology* 35 (1999): 4–28.

Steele, C. M. "A threat in the air: How stereotypes shape intellectual identity and performance." *American Psychologist* 52 (1997): 613–629.

Steele, C. M., and J. Aronson. "Stereotype threat and the intellectual test performance of African Americans." *Journal of Personality and Social Psychology* 69 (1995): 797–811.

20. Burgess, D. J., A. Joseph, et al. "Does stereotype threat affect women in academic medicine?" *Academic Medicine* 87(4) (2012): 506–512.

Brodish, A. B., and P. G. Devine. "The role of performance-avoidance goals and worry in mediating the relationship between stereotype threat and performance." *Journal of Experimental Social Psychology* 45 (2009): 180–185.

Davies, P. G., S. J. Spencer, and C. M. Steele. "Clearing the air: Identity safety moderates the effects of stereotype threat on women's leadership aspirations." *Journal of Personality and Social Psychology* 88 (2005): 276–287.

Croizet, J.-C., G. Després, et al. "Stereotype threat undermines intellectual performance by triggering a disruptive mental load." *Personality and Social Psychology Bulletin* 30 (2004): 721–731.

Keller, J., and D. Dauenheimer. "Stereotype threat in the classroom: Dejection mediates the disrupting threat effect on women's math performance." *Personality and Social Psychology Bulletin* 29 (2003): 371–381.

Schmader, T., and M. Johns. "Converging evidence that stereotype threat reduces working memory capacity." *Journal of Personality and Social Psychology* 85(3) (2003): 440–452.

Steele, C. M. "A threat in the air: How stereotypes shape intellectual identity and performance." *American Psychologist* 52 (1997): 613–629.

21. Krendl, A. C., J. A. Richeson, et al. "The negative consequences of threat: A functional magnetic resonance imaging investigation of the neural mechanisms underlying women's underperformance in math." *Psychological Science* 19 (2008): 168–175.

22. Plant, E. A., and P. G. Devine. "The antecedents and implications of interracial anxiety." *Personality and Social Psychology Bulletin* 29 (2003): 790–801.

23. Dovidio, J. F. "On the nature of contemporary prejudice: The third wave." *Journal of Social Issues* 57 (2001): 829–849.

24. Biernat, M., M. Manis, and T. E. Nelson. "Stereotypes and standards of judgment." *Journal of Personality and Social Psychology* 60 (1991): 485–499.

25. Biernat, M., and M. Manis. "Shifting standards and stereotype-based judgments." *Journal of Personality and Social Psychology* 66 (1994): 5–20.

26. Dovidio, J. F., and S. L. Gaertner. "Aversive racism and selection decisions: 1989 and 1999." *Psychological Science* 11 (2000): 315–319.

27. Bielby, W.T., and J. N. Baron. "Men and women at work: Sex segregation and statistical discrimination." *American Journal of Sociology* 91 (1986): 759–799.

28. Biernat, M., and M. Manis. "Shifting standards and stereotype-based judgments." *Journal of Personality and Social Psychology* 66 (1994): 5–20.

29. Deaux, K., and T. Emswiller. "Explanations of successful performance on sex-linked tasks: What is skill for the male is luck for the female." *Journal of Personality and Social Psychology* 29 (1974): 80–85.

30. Martell, R. F. "Sex bias at work: The effects of attentional and memory demands on performance ratings of men and women." *Journal of Applied Social Psychology* 21 (1991): 1939–1960.

31. Eagly, A. H., and S. Sczesny (2009). "Stereotypes about women, men, and leaders: Have times changed?" In *The Glass Ceiling in the 21st Century: Understanding Barriers to Gender Equality*, edited by M. Barreto, M. K. Ryan, and M. T. Schmitt, 21–47. Washington, DC: American Psychological Association.

   Eagly, A. H., and A. M. Koenig (2008). "Gender prejudice : On the risks of occupying incongruent roles." In *Beyond Common Sense: Psychological Science in the Courtroom*, edited by E. Borgida and S. T. Fiske, 63–81. Malden, MA: Blackwell Publishing.

   Heilman, M. E., A. S. Wallen, et al. "Penalties for success: Reactions to women who succeed at male gender-typed tasks." *Journal of Applied Psychology* 89 (2004): 416–427.

   Ridgeway, C. L. "Gender, status, and leadership." *Journal of Social Issues* 57 (2001): 637–655.

32. Dovidio, J. F., et al. "On the nature of prejudice: Automatic and controlled processes." *Journal of Experimental Social Psychology* 33 (1997): 510–540.

33. Trix, F., and C. Psenka. "Exploring the color of glass: Letters of recommendation for female and male medical faculty." *Discourse & Society* 14 (2003): 191–220.

34. Steinpreis, R. E., K. A. Anders, and D. Ritzke. "The impact of gender on the review of the curricula vitae of job applicants and tenure candidates: A national empirical study." *Sex Roles* 41 (1999): 509–528.

35. Wennerås, C., and A. Wold. "Nepotism and sexism in peer-review." *Nature* 387 (1997): 341–343.

36. Budden, A. E., et al. "Double-blind review favours increased representation of female authors." *Trends in Ecology & Evolution* 23 (2008): 4–6.

37. Johns, M., T. Schmader, and A. Martens. "Knowing is half the battle: Teaching stereotype threat as a means of improving women's math performance." *Psychological Science* 16 (2005): 175–179.

38. Good, J. J., J. A. Woodzicka, and L. C. Wingfield. "The effects of gender stereotypic and counter-stereotypic textbook images on science performance." *Journal of Social Psychology* 150 (2010): 132–147.

READING

Quotes

Hurtado, Sylvia. "Linking diversity with the educational and civic missions of higher education." *Review of Higher Education* 30 (2007): 186.

Uhlmann, Eric Luis, and Geoffrey L. Cohen. "'I think it, therefore it's true': Effects of self-perceived objectivity on hiring discrimination." *Organizational Behavior and Human Decision Processes* 104 (2007): 208.

Prepared for WISELI by Eve Fine and Jo Handelsman

Thanks to Molly Carnes, Jennifer Sheridan, Amy Wendt, Linda Baier Manwell, Brad Kerr, and Christine Calderwood for their suggestions.

READING

Assessing Understanding

Introduction

Determining whether mentees understand core concepts about the research they are doing is critical in a productive mentoring relationship, but is surprisingly difficult to do. Developing strategies to assess how well mentees understand the purpose of their experiments, the principles underpinning experimental techniques, and the context of their work in the lab and field is an important part of becoming an effective mentor. Moreover, it is important for mentors to be able to identify the causes for confusion and strategies to address misunderstandings.

Learning Objectives

Mentors will have the knowledge and skills to

1. **Assess their mentee's understanding of core concepts and processes and ability to develop and conduct a research project, analyze data, and present results**
2. **Identify reasons for a lack of understanding, including expert-novice differences**
3. **Use multiple strategies to enhance mentee understanding across diverse disciplinary perspectives**

Overview of Activities for the "Assessing Understanding" Session

	Learning Objectives	Core Activities
1	Assess their mentee's understanding of core concepts and processes and ability to develop and conduct a research project, analyze data, and present results	Mentors read and discuss the scenarios (Activity #1)
2	Identify reasons for a lack of understanding, including expert-novice differences	Mentors brainstorm reasons for a lack of understanding (Activity #2)
3	Use multiple strategies to enhance mentee understanding across diverse disciplinary perspectives	Mentors share strategies to enhance understanding (Activity #3)

FACILITATION GUIDE

Recommended Session for Assessing Understanding (60 minutes)

Materials Needed for the Session:

- Table tents and markers
- Chalkboard, whiteboard, or flip chart
- Handouts:
  - Copies of introduction and learning objectives for Assessing Understanding (page 83)
  - Copies of "Assessing Understanding" scenarios (page 87)
  - Copies of Mentoring Tools (pages 88–92)

Overview (5 min)

TELL ▶ Review the introduction and learning objectives for the session. Be clear that this session is about assessing a mentee's understanding of research concepts and processes. While understanding other factors that impact mentor-mentee relationships is important, keep the focus on research.

Objective 1: Assess their mentee's understanding of core concepts and processes and ability to develop and conduct a research project, analyze data, and present results (25 min)

ACTIVITY #1: Scenarios

READ ▶ (2 min) Distribute the "Assessing Understanding" scenarios. Assign each small group one of the scenarios. Let participants read their assigned scenario individually.

DISCUSS ▶ (10 min) Have each small group discuss its assigned scenario among the group's participants. Each group should come up with three specific approaches to avoiding or resolving the described situation.

ASK ▶ (13 min) Ask each group to share two approaches with the entire group. You may want to record the ideas generated in this discussion on the whiteboard or flip chart. Here are some additional questions for discussion:

1. How do know if your mentee understands something?
2. How can you help your mentees accurately assess their own understanding?
3. How often should a mentor check in with their mentee about their understanding?
4. Mentors can make assumptions about a mentee who does not understand. They may think the mentee is blowing off the work. How do you determine the difference between a mentee not understanding something and a mentee not trying?
5. How can you deal with the fact that a mentee may be embarrassed by a lack of understanding?
6. How can mentors balance promoting independence with confirming understanding?
7. How can you tell the difference between a miscommunication and a true lack of understanding?

Objective 2: Identify reasons for a lack of understanding, including expert-novice differences (15 min)

ACTIVITY #2: Follow-Up Discussion (15 min)

DISCUSS ▶ Pose the following questions to the group. You may want to record the ideas generated in this discussion on the whiteboard or flip chart.

- What reasons might there be for a mentee having difficulty understanding?
- We all unconsciously make assumptions about ability and level of understanding based on other cues and factors, such as race, ethnicity, gender, English fluency, prior experience and background, the types of questions someone asks, and so on. How can you acknowledge those assumptions and remain open-minded?

NOTE ▶ Some of the reasons that may arise include differing backgrounds, different modes of communication, misunderstandings regarding the level of understanding that is expected, cultural differences, disciplinary differences, and so on.

NOTE ▶ You may want to ask mentors to consider the difference between an expert perspective and a novice perspective; for example, as an expert, there are many steps in an explanation you may leave out as they are second nature or it is hard to remember what it was like to be a novice. When you see a master chef cooking, it looks easy; however, when you try to make the same dish yourself, you realize that there are many steps that have been left out of the explanation.

Objective 3: Use multiple strategies to enhance mentee understanding across diverse disciplinary perspectives (15 min)

ACTIVITY #3: Identifying Strategies to Enhance Understanding (15 min)

ASK ▶ Have mentors generate a list of strategies that can be used to assess their mentee's understanding as well as ways to remedy a lack of understanding. Ask mentors to consider strategies that can be used in face-to-face meetings, over email, through written reports, and so on. You may want to record the ideas generated in this discussion on the whiteboard or flip chart. Strategies you can add to the list include the following:

- Taking a minute to consider their assumptions about what mentees know or do not know

- Taking time to remember what it was like not to understand something before they became an expert
- Writing out an explanation and asking one of their peers from outside the discipline to identify all the terms they do not understand
- Asking their mentee to explain something so they can assess the mentee's understanding (this could be done verbally during a meeting, or afterward with the main points briefly summarized via email)
- Asking their mentee to explain something to another trainee
- Asking their mentee to organize information in a flowchart, diagram, or concept map

NOTE ▶ You may want to ask participants the following: How do they know when they are qualified to assess a mentee's understanding? What do they do if they are not an expert in all aspects of a mentee's research program, such as when they are a secondary mentor?

Postsession Assignments

TELL ▶ Have participants return to their compacts (if applicable) and make changes based on their reflections on assessing understanding.

TELL ▶ Suggest that everyone present a mentoring challenge to their own adviser/research mentor or another faculty member they respect and discuss possible solutions. Advise participants to be prepared to share highlights from their discussion at the next session.

Mentoring Tools*

The following mentoring tools are located at the end of this chapter:

- Your Research Group's Focus
- Scientific Article Worksheet
- Research Project Outline & Science Abstract

These tools are intended to help students understand and communicate their individual research projects and scientific articles.

* These tools are adapted from Branchaw, J. L., Pfund, C., and Rediske, R. (2010), *Entering Research: A Facilitator's Manual: Workshops for Students Beginning Research in Science*, W.H. Freeman & Company.

FACILITATION GUIDE

"Assessing Understanding" Scenarios

Scenario A: Revealing Abstract

You have just spent the last month working intensively with your new undergraduate mentee. You have given her multiple papers to read and have had several discussions about your research. In addition, she has engaged in several different aspects of an ongoing project over the last month. She is hardworking and seems to understand the group's research, and things seem to be going well. On Monday morning, she hands you a draft of the introduction section for a possible senior thesis project. After reading through the draft, you are forced to conclude that she does not understand the work of your lab.

What can you do to address this situation? How can you avoid this situation in the future?
Come up with at least three strategies for avoiding this situation.

Scenario B: It Seemed So Clear When You Explained It

You have recently explained a complicated technique to your mentee. While you were explaining, he nodded the entire time as if he understood every word you were saying. When you were finished with your explanation, you asked him if he had any questions. He said no. Just to make sure, you asked him if everything was clear. He said yes. Three days later you asked the mentee how his work using this technique was going, and he told you he hadn't started because he did not understand the technique.

What can you do in the future to make sure your mentee understands what you are saying?
Come up with at least three strategies for assessing your mentee's understanding.

Scenario C: It Just Didn't Work

I have a really promising mentee who's doing well in all of his upper-division major courses. When we work through experiments together, he knows all the right techniques, but he doesn't seem to be able to get experiments to work right when he's by himself. I'm trying to help him figure out what's happening in his failed experiments, but our conversations all seem to go like this:

"So what happened with your reaction?"

"It didn't work."

"What happened?"

"Nothing. It just didn't work."

"What do you think went wrong?"

"I don't know. But I tried it twice and it didn't work either time."

We're both getting a little frustrated with the lack of progress, and I've noticed that he's started spending less time in the lab.

Suggest approaches to get things back onto the right track.
Come up with at least three strategies for dealing with this situation.

Mentoring Tool

Your Research Group's Focus[*]

> **Objective:** Students will learn about their group's overall research goals and begin to understand how their individual research project will contribute to these goals.

Write one paragraph, *in your own words*, describing the focus of your group's research. Be sure to include the group's major research questions or hypotheses, the types of techniques they use to investigate these questions, and what area(s) of this work are most interesting to you.

To share this with the rest of the research team, students can give "chalk talks" (informal oral presentations) at lab meeting in which they draw a diagram on the board to explain the research.

---

[*] Adapted from Branchaw, J. L., Pfund, C., and Rediske, R. (2010), *Entering Research: A Facilitator's Manual: Workshops for Students Beginning Research in Science*, W.H. Freeman & Company.

Mentoring Tool

Scientific Article Worksheet[*]

> **Objective:** Students will learn strategies to effectively and efficiently read scientific articles.

Title:

Authors:

Journal:

Year:

The Basics:

1. What hypothesis or research question does the paper address?

2. What experiments were done to test the hypothesis or investigate the research question?

3. What are the major conclusions?

4. What evidence supports each of the conclusions?

The Critique:

1. Is the paper well written? How do you know?

2. Do the conclusions seem logical given the data presented? Why or why not?

3. Why are the conclusions important?

4. What were the best aspects of the research presented, and how could it be improved?

---

[*]From Branchaw, J. L., Pfund, C., and Rediske, R. (2010), *Entering Research: A Facilitator's Manual: Workshops for Students Beginning Research in Science*, W.H. Freeman & Company.

Additional Resources:

1. What are the basic concepts that you need to know to understand the science presented in your paper?

2. Identify a chapter/section in a textbook that outlines these basic concepts. Is reading this helpful to your understanding?

3. What other information or resources would help you better understand the paper?

Further Questions:

Write *at least* five comments or questions about the article to discuss with your mentor.

1.

2.

3.

4.

5.

Mentoring Tool

Research Project Outline & Science Abstract*

> **Objective:** Students will summarize their research project for their peers and write a scientific abstract.

Research Group's Focus:

Research Project Title:

Introduction/Background:

Identify and summarize the key background information needed to understand your research project. Write these pieces of information as a *bulleted list of statements*. Your hypothesis or research question should follow from this information.

-
-
-

Hypothesis or Research Question:

Relevance and Implications of Your Research Project:

Why is your research important? What may be the potential implications of your results? How will your project benefit basic research, human health, or development of a commercial product?

Experimental Design and Potential Results:

Outline the experiments you will do to test your hypothesis. For each experiment, explain

1. the technique(s) that will be used and the reason(s) for selecting that technique.

---

*From Branchaw, J. L., Pfund, C., and Rediske, R. (2010), *Entering Research: A Facilitator's Manual: Workshops for Students Beginning Research in Science*, W.H. Freeman & Company.

2. the type of data that will be collected and why this type of data will inform the hypothesis.

3. all the potential results and whether each would support, or not support, your hypothesis. Draw what the predicted results will look like, if applicable (e.g., gel, microscope image, data table, or graph).

Timeline:

Outline a weekly or monthly timeline for your project. Be sure to refer to each of the proposed experiments (or parts of the experiments), allow time for analysis of data, and allow time for the preparation of a presentation of the data (e.g., poster or oral presentation).

Abstract:

Synthesize the core information in your outline and write a scientific abstract of 200 words or less.

CHAPTER

7

Fostering Independence

Introduction

An important goal in any mentoring relationship is helping the mentee become independent; yet defining what an independent mentee knows and can do is often not articulated by the mentor or the mentee. Defining what independence looks like and developing skills to foster independence are important to becoming an effective mentor. Defining independence becomes increasingly complex in the context of team science.

Learning Objectives

Mentors will have the knowledge and skills to

1. **Define independence, its core elements, and how those elements change over the course of a mentoring relationship**
2. **Employ various strategies to build their mentee's confidence, establish trust, and foster independence**
3. **Create an environment in which mentees can achieve goals**

Overview of Activities for the "Fostering Independence" Session

	Learning Objectives	Core Activities
1	Define independence, its core elements, and how those elements change over the course of a mentoring relationship	Define skills and behaviors that demonstrate independence (Activity #1)
2	Employ various strategies to build their mentee's confidence, establish trust, and foster independence	Read and discuss the case study "Ready Mentee" (Activity #2)
3	Create an environment in which mentees can achieve goals	Read and discuss the case study "The Slow Writer" (Activity #3)

FACILITATION GUIDE

Recommended Session for Fostering Independence (60 minutes)

Materials Needed for the Session:

- Table tents and markers
- Chalkboard, whiteboard, or flip chart
- Handouts:
  - Copies of introduction and learning objectives for Fostering Independence (page 93)
  - Copies of the case study "Ready Mentee" (page 97)
  - Copies of the case study "The Slow Writer" (page 97)
  - Copies of the reading "Mentoring Research Writers" (pages 98–104)

Introductions (10 min)

ASK ▶ To follow up on the last session's assignment, have each participant in the group answer the following questions:
- What case did you present to your adviser/research mentor or other faculty member?
- What did that person say?

TELL ▶ Review the introduction and learning objectives for this session.

Objective 1: Define independence, its core elements, and how those elements change over the course of a mentoring relationship (20 min)

ACTIVITY #1: Defining Independence (10 min)

ASK ▶ Have participants describe their definition of independence. What does independence look like for an undergraduate researcher? What should the mentee know and be able to do by the time they leave the mentor's lab?

NOTE ▶ You may want to record the ideas generated in this discussion on the whiteboard or flip chart. Here are some elements of independence:

- Basic knowledge of discipline
- Ability to critically read the literature and find answers to questions through literature searches and consulting experts
- Ability to present a poster on their work at a national meeting
- Ability to design an experiment

DISCUSS ▶ (10 min) Engage the entire group in a discussion of the following questions:

1. How can you tell whether a certain level of independence is achieved? What does independent thinking and behavior look like?
2. Do mentees know what level of independence they are expected to achieve? Do they understand that this will change throughout their careers?
3. How well aligned are your estimates and those of your mentee regarding their level of independence?
4. How can a mentee work both as an independent researcher and as a member of a research team?

FOLLOW-UP ACTIVITY: Draw a timeline for establishing independence and discuss it with your mentee to see if it aligns with their expectations. You may consider adding this timeline to your compact (if applicable).

Objective 2: Employ various strategies to build their mentee's confidence, establish trust, and foster independence (15 min)

ACTIVITY #2: Case Study: Ready Mentee

READ ▶ (2 min) Distribute the case study and ask participants to read the case to themselves.

DISCUSS ▶ (13 min) Discuss the case with the entire group. You may want to record the ideas generated in this discussion on the whiteboard or flip chart. Use the guiding questions following the case study to facilitate discussion. Additional questions include the following:

- How do you convey to your mentee that mistakes are part of learning science?
  - How do you judge which and how many mistakes are acceptable?
  - How important is it for a mentee to make mistakes to become independent?
- Can you give a mentee too much independence? How would you know?
- How do you determine the underlying reason for a mentee's need to ask for constant advice? (Consider personality, forgetfulness, fear of making mistakes, cultural differences, or lack of experience, which may lead to such behavior.)
- How can you determine whether assumptions about an undergraduate researcher's background, ethnicity, gender, or some other factor are influencing your assessment of their research skills?
- What are possible consequences of talking negatively about an undergraduate researcher to other researchers in the lab?

Objective 3: Create an environment in which mentees can achieve goals (15 min)

ACTIVITY #3: Case Study: The Slow Writer

READ ▶ (2 min) Distribute the case study and ask one participant to read the case aloud.

DISCUSS ▶ (13 min) Engage the entire group in discussion. You may want to record the ideas generated in this discussion on the whiteboard or flip chart. Use the guiding questions following the case study. Here are some additional questions as well:

- How do you establish and communicate your expectations of your mentee?
- What do you do when your mentee repeatedly does not meet your expectations?
- What are strategies for uncovering the unspoken expectations mentees and mentors may have about issues such as authorship, hierarchy, letters of recommendation, and so on?
- How can you help a mentee navigate the different expectations articulated by multiple mentors?

NOTE ▶ You may wish to refer mentors to the reading "Mentoring Research Writers," included in the handouts for this session.

Postsession Assignment

TELL ▶ Have participants look up a general ethics statement for their field published by the professional society most closely associated with their research. If they can't find one, tell them to ask a senior colleague about it or perhaps ask their society why they don't have one. As a last resort, suggest that they start drafting one on their own.

TELL ▶ Have everyone read "Mentoring Research Writers."

FACILITATION GUIDE

Case Study: Ready Mentee

An experienced undergraduate researcher was constantly seeking input from his mentor on minor details regarding his project. Though he had regular meetings scheduled with his mentor, he would bombard her with several emails daily or seek her out any time she was around, even if it meant interrupting her work or a meeting that was in progress. It was often the case that he was revisiting topics that had already been discussed. This was becoming increasingly frustrating for the mentor, since she knew the student was capable of independent work (having demonstrated this during times she was less available). The mentor vented her frustration to at least one other lab member and wondered what to do.

Guiding Questions for Discussion:

1. What are the main themes raised in this case study?

2. What should the mentor do?

3. How do you determine how much independence a mentee is ready for?

Case Study: The Slow Writer

A senior undergraduate student in my group is adept at performing experiments and analyzing data but is a very slow writer. Last semester, I set multiple deadlines that this student missed, while another student in my group wrote an entire senior thesis chapter, coauthored a paper, and did experiments. Over winter break, the slow writer had a breakthrough and produced a fairly reasonable draft of her senior thesis. To avoid delays in publications, I have begun revising this draft, rather than simply providing comments. However, to graduate, I realize that she must write the thesis herself, as well as the section of a manuscript she wants to coauthor. Setting deadlines for detailed outlines, thesis sections, and figures hasn't worked. Communicating the importance of writing to the scientific endeavor hasn't worked. Encouragement hasn't worked. Veiled threats don't seem professional. Other than being patient, what should I do?

Guiding Questions for Discussion:

1. What are the main themes raised in this case study?

2. What could have been done to avoid this situation? What should the mentor do now? What should the mentee do now?

3. How do you find out what expectations your mentee has of you and for their research experience?

Mentoring Research Writers

by Bradley Hughes[*]

Recognizing the Power of Writing as a Component of the Research Process

As a mentor you have a great opportunity to encourage your trainees to set high goals for their research writing and to help them achieve those goals. You should recognize, in fact, that you have a serious responsibility to motivate and to help researchers-in-training become excellent writers. Why should you and your trainees make writing a priority? The answer is clear to all experienced researchers: researchers earn their living and develop their careers *through the writing they do*—writing proposals to fund research, writing conference abstracts and posters and papers to disseminate new knowledge and to influence future research and the shape of their fields, documenting their research methods and findings, writing reviews of literature, writing reviews of colleagues' manuscripts, and writing letters of recommendation. Writing pervades the research process, and successful researchers spend a significant amount of their time planning, drafting, and revising complex forms of writing. Experienced researchers also know that writing is not just a way to communicate completed findings and polished arguments: writing is actually a powerful form of thinking and learning, one that clarifies thought and makes analyses and arguments more precise.

Acknowledging the Complexity of Research Writing

In order to appreciate the complexity of research writing and to guide new researchers, mentors need to understand that writing is a highly situated practice—that is, it is not a generic, general skill. Successful researchers need to achieve very specific purposes and speak persuasively to particular groups of readers. What is valued in writing and what is conventional and effective in writing varies across particular scientific communities and even within particular communities of researchers.

As researchers transition from writing within particular disciplines or professions to new ones, they often struggle to write successfully, even if they had success in previous writing situations. Given how varied purposes and audiences are for advanced research writing, as a research mentor, you should have intentional conversations about research writing with your mentees—working on and talking about writing are natural and important parts of training programs, and you should not expect new biomedical researchers to be accomplished writers from the start. Becoming an excellent research writer takes time, effort, and dedicated, consistent mentoring.

Mentors should also remember that researchers-in-training, like all students, bring varied literacy backgrounds to each new writing challenge. Some of your research trainees will have done lots of writing and reading, been held to high standards for written communication, and learned to receive and give critical feedback on writing. Others may feel that their intellectual strengths lie in quantitative rather than verbal areas. Some may have great strengths in oral communication rather than academic writing. Others may be multilingual writers, who are very skilled communicators in their first or second languages and who have great cross-cultural linguistic knowledge, but less experience writing and reading English. Some multilingual writers may have internalized organizational structures or styles for academic writing from their first language that are at odds with standard patterns in English. Still other writers may have a tenuous grasp on the subject that they are writing about,

[*]Director, The Writing Center and Writing Across the Curriculum, Department of English, University of Wisconsin–Madison

and their conceptual struggles may manifest themselves in their writing. At the same time, many researchers find writing difficult and as a consequence avoid writing, procrastinate, and eventually end up in stressful time crunches that reinforce their dislike for writing.

Key Principles in Mentoring Writers

1. Signal from the very start and reinforce frequently that excellent writing is a high priority for you, for your research group, and for all successful researchers.

2. Figure out what your mentees already know about research writing and find ways to help them learn what they need to learn.

3. Work collaboratively with your research mentees to motivate them to write every week, sometimes every day.

4. Talk with your mentees regularly about their writing—analyzing successful examples, planning new pieces of writing, brainstorming, kicking ideas around, discussing drafts, and planning revisions.

5. Schedule meetings to plan and work on drafts. Make discussions of in-progress writing part of the culture and rhythm of your research group.

6. Give clear, specific, encouraging feedback. Start first with global concerns and then move on to more local, smaller concerns.

7. Be sure your feedback identifies strengths and potential as well as problems.

8. Honor and celebrate successful research writing within your research group.

Given what varied experiences and strengths researchers-in-training may bring, you should ask your mentees about their previous experience and about their perceived strengths and areas for improvement. Acknowledge that research writing is always hard work, especially when researchers are learning to write in a new field or in a new genre, when they are making arguments that are more complex than they have made before, or when they're not sure what their findings mean or what is interesting or important in their findings. For these reasons, research writers need their mentors to be patient and encouraging as well as critical. And above all, mentors need to *normalize revision*; revision is a normal and crucial part of writing, not a sign that a writer has failed because she or he did not achieve perfection in an early draft. Research shows that experienced, successful writers spend a lot of time revising their work.

Writing is hard work and time-consuming for mentees. Let's face it—helping mentees learn to become strong research writers is hard work and time-consuming for you as a mentor. Although the recommendations that follow should make the time you spend on mentoring more successful and effective for you and for the writers you are mentoring, there are no shortcuts. Reading drafts carefully and critically and charitably; discerning what is and what is not working well in a draft; giving clear, specific, helpful, and encouraging feedback; reading yet another draft; meeting to talk through your feedback and the writer's plan for revision—these critical tasks will always require concentration and time. But they are what every writer needs in order to learn and to improve—to become the strongest possible research writer they can be and to launch their research career.

Here are some specific strategies, drawn from research and practice, for mentors to try.

Before the First Draft

Find ways to signal that writing is crucial to research in your field and that mentoring researchers to become strong writers is a high priority for you and for your research group. When, for example, a prospective researcher interviews with you, talk about writing and your commitment to mentoring writing. If you use some form of written expectations, such as a mentoring compact, you might consider including a section for your mentees on writing. Create a culture within your group of sharing and discussing drafts and of sharing and celebrating successful writing. In your meetings or discussions, always find time to talk about writing—even long before it is time to begin writing.

Talk with trainees about their writing processes, and yours. You might read and discuss writing resources, which offer valuable advice about establishing good habits for academic writing. You might also want to share some drafts of your own research writing in progress, seeking feedback from your mentees—learning to give constructive, critical feedback helps writers grow, and sharing your drafts will give you valuable feedback and model the drafting, critique, and revision process that you are trying to teach.

Recognize that *talk* is a crucial part of writing. Be sure that you are talking regularly with trainees about their writing in progress. Your mentoring discussions about research questions, methods, literature, and results are all critical for helping a newer researcher figure out how they will explain their research project in research publications, in funding proposals, in presentations, and in interviews. In discussions, ask questions that point toward future writing, such as

> "How are you thinking about organizing your literature review?"
>
> "How might you phrase that as a research question?"
>
> "In your results, what's new? What's most significant?"

These kinds of questions and many others help researchers clarify their thoughts through talk and help them prepare for writing. And by your choice of questions, you are helping reinforce the key principles of scientific research and helping researchers imagine the audiences for whom they will be writing.

Your trainees will benefit if you ask them to prepare and discuss the main information and arguments in their papers. Researchers benefit from having to organize information in a logical outline and giving colleagues a chance to ask questions and offer advice *before* investing hours and hours in drafting sentences and paragraphs. You might ask them to prepare and discuss informally, with you and with peers, a few PowerPoint slides outlining the main information and arguments they hope to include in their paper. Another good reason to invest time up-front clarifying key ideas and arguments: if you and your mentee do *not* clarify and agree on the main points and arguments for the paper early in the process of writing, don't be surprised if your mentee is reluctant to make major changes after she or he has invested all the time that it takes to write a full draft.

New research writers need to develop a robust understanding of the *genres* commonly written by researchers in their discipline. Strong, successful research writers can take an aerial view of a document and can talk intentionally about the purpose of a particular piece of writing and about the choices authors have made about the content and organization for a given genre. Mentors should work systematically with mentees to identify and to analyze the key genres (or kinds of writing) in relevant fields or subfields, looking at what a particular kind of writing accomplishes and how it is tailored to a particular audience. For each key genre, mentors should first explore mentees' experience and understanding about that genre. As you have these discussions, you might want to ask trainees to analyze, together with you, the different kinds of articles in major journals in your field. In talking about genre, try to focus not on surface features of a genre (e.g., the citation system) but aim to develop—in yourself as a mentor and in your mentees—an ability to talk about the rhetoric of each genre; that is, the purpose of that genre, its audience, and its persuasive elements. For example, talk systematically about which questions get answered in the introduction, in the literature review, in the methods, in the results, and in the discussion sections. How is information organized *within* a particular section (such as the results section)? How much detail do authors give? What do the authors assume about the knowledge their readers already have about the topic under study?

Engage in "prewriting." Before your mentee begins drafting a proposal or research report, use your conversations to help your mentee plan and do what is called "prewriting." You can use your time—and your mentee's time—wisely by doing some explicit planning of a paper before your mentee starts actually drafting sections of it. Through collaborative talk and questions, you can help an author clarify the purpose of a piece of writing, central research questions, a plan, an outline, lists of main points, and the logic of an argument. Moreover, you can capture good ideas, plans, and important language—the mentee's and yours—by writing them down often as they emerge in these conversations. Your conversation and interest and encouragement also provide crucial motivation for doing the hard work of starting a writing project. And by correcting major misconceptions at this stage, you're helping writers, rather than waiting for a writer to invest countless hours in writing a full draft that may be misguided in some fundamental ways.

Set intermediate deadlines for portions of a draft, and insist that mentees meet those deadlines. Less experienced research writers need to write a partial draft long before they think they are ready to write, in order to give mentors a chance to give formative feedback and in order to give mentees plenty of time to revise. Early drafts, tough but encouraging critical feedback, and lots of revisions—these are what produce strong thinking and strong scientific writing. You might consider scheduling a weekly draft discussion for all lab members, with different members scheduled to share their work each week. It is natural for busy postdocs or graduate students to fall behind with deadlines, and of course mentors should be understanding and flexible, but you are not doing your mentees a favor if you allow them to delay writing for too long. Be sure your expectations for writing are clear and that the mentee understands the consequences of falling behind in writing given the number of publications they are expected to produce while working with you.

Ask your trainees to include a cover sheet with each draft. Each time your mentee provides you with a draft of their writing it should be accompanied by a cover sheet, which can orient you as a reader. This cover sheet might include relevant questions, such as

- What is this draft?
- Who is the intended audience?
- How is it organized?
- What are your main points?
- What do you think is working well? What are you pleased with?
- What would you especially like me to focus on as I read, or what would you like my help with?

Answers to these questions can guide your reading, and you will be able to use your time more effectively and be sure to respond to the writer's needs. Learning to reflect critically on their own writing is valuable for writers as well; experienced writers can talk effectively about their writing, can offer an aerial view of a draft, and can ask readers for particular kinds of help.

Giving Feedback and Guiding Revisions on Drafts

Encourage mentees to welcome criticism and advice about their writing. Before you ever give specific feedback on a draft, find comfortable ways to ask your mentees about their experience receiving feedback on drafts and about their feelings about feedback and criticism. Talk about your own feelings about advice and criticism and encourage your mentee to welcome and consider all feedback, to ask for clarification during an in-person conversation, and to feel comfortable choosing not to accept some advice but justifying that choice. Explain that the strongest, most successful writers seek out tough, critical readers while their writing is still changeable.

Explain your approach to feedback and contextualize your comments. For example, if you have commented only on big ideas or the next steps you are suggesting, be sure to tell that to the writer. Otherwise, it is easy for a writer to assume that because you have not commented on something that means there are no problems with it. If you commented on local concerns only in one section but similar problems continue in other parts of the draft where you did not comment, be sure to explain this lack of feedback that so that writers do not have to guess what it means.

Focus first on global concerns before local concerns. In your reading, in your comments, and in your conversations with the writer, focus first on whether the big picture is working well by addressing *global, high-level concerns* like these:

- Is the central research question clear?
- Is the significance of the research clear and persuasive?
- Is the progression of ideas and arguments logical?
- Does the writer demonstrate a clear understanding of the major concepts under study?
- Does the review of literature emphasize the most important ideas?
- Are findings clearly explained and easy to grasp—in figures and graphs as well as in the text?
- Are ideas thoroughly explained?
- Is the discussion focused on the most important points?

Later in the process of writing and revising, when the big stuff is working pretty well, narrow your focus and the writer's to more *local concerns* like these:

- Are there effective transitions between sections?
- How can the style be improved?
- Where do sentence or word problems interfere with the writer's ability to communicate clearly?
- Are there any grammatical errors?
- How can the word choice be improved?
- Are there punctuation errors?
- Are there proofreading mistakes?

Why is it important to start our feedback with global concerns? First, it is just a matter of efficiency—you have limited time to give feedback and your trainees have limited time to revise, so there is not much point to your commenting on small edits and not much point to the writer's making small edits when the writer needs to make larger changes. Second, research shows that less experienced writers are often confused by what faculty and mentors want them to concentrate on in their writing and in their revisions. They may think, for example, that correcting semicolon mistakes or rephrasing part of a sentence is as important as clarifying the logic of their discussion or anticipating and addressing counterarguments or emphasizing some ideas and subordinating others. And mentor comments on their writing too often lead writers to make only superficial revisions to words and sentences, overlooking larger conceptual, rhetorical, and structural revisions that would most improve a paper. By starting your feedback with global concerns, mentees get clear guidance from you about how to strengthen their ideas, their analyses, and their arguments, so that they have papers worth editing and polishing. *Then* you can turn your attention—and your trainees' attention—to improving sentences, words, and punctuation.

Identify strengths and potential in a draft, teach from success, and offer encouragement. In your comments, instead of jumping right into what's wrong or needs improving, try starting with what you see as the specific strengths in a draft, what's promising, and what's working well. And it's important to make some of your praise specific, as specific as some of your criticism. So instead of saying "Good start," or just "Good," try identifying what in particular is working well in a draft. This does not mean to offer false or insincere praise, but writers need to know what they are doing well and they need to see you as a reader who is genuinely interested in what they have to say and eager for them to succeed, rather than seeing you only as an error hunter. Teaching or coaching for success means if a writer has done something well in one section of a draft (if, for example, their topic sentences orient a reader well to the topic and main point of a paragraph) but not in another section, you can encourage the writer to do what they have already done well elsewhere.

Be direct and clear in your request for revisions. When giving feedback, indicate in specific terms how much work remains to be done. For example, "This will need a fair amount of revision in order to clarify your key research questions and to report your key findings effectively. As you revise, here are my key suggestions: (1) . . . ; (2) . . . ; (3) . . . ." Or "After you've worked on focusing the literature review around just a few central concepts, you'll need to do some substantial editing to clarify sentences. I've shown the kinds of edits in the first paragraph of the lit review, but the rest of the draft needs that same kind of editing." You can be clear and constructive in your feedback, even if you are delivering bad news, but you are not doing a writer any favors if you hide or sugarcoat how much work remains to be done.

Ask writers to document their revisions. When you're reviewing a revised version of something you've read before, ask the writer to attach a cover sheet (described on pages 101–102) explaining the major changes they've made since you last read it. Asking trainees to do this signals that you expect them to make major revisions before you read something again. This kind of cover letter resembles what you would write in a cover letter or email with a revised manuscript if you received a "revise and resubmit" decision from a journal editor. In addition, you might want to ask the trainee to use "track changes" so that you can focus your reading on what's changed.

Close your comments with some encouragement and a look forward. Be sure to include notes of encouragement and expectation with your feedback. For example, you might say, "Looking forward to reading the next draft of this," or "Looking forward to seeing this in print soon!" or "Looking forward to meeting on Thursday to talk through your plans for revising."

Within your research group, create a culture that celebrates important milestones in writing. Acknowledge and celebrate proposals and manuscripts when they are submitted, when revisions are completed, grants funded, publications accepted, and publications appear.

Mentors play a critical role in helping researchers-in-training become excellent, independent writers. Be sure to set the bar high for your trainees' thinking, research, and writing and then provide them with support to meet those expectations. If at any point you feel that a mentee requires additional feedback and support, seek out local resources and encourage your mentee to take advantage of them.

8

Cultivating Ethical Behavior

Introduction

Mentors play an important role in both teaching and modeling ethical behavior. There are ethical issues centering on the science itself—how to conduct, report, and write scientific studies—as well as relationships between mentors and their mentees. Most relationships in science establish a power dynamic, and it is a mentor's responsibility to learn how to manage their power. Reflecting upon and discussing ethical behavior is an important part of becoming an effective mentor.

Learning Objectives

Mentors will have the knowledge and skills to

1. **Articulate ethical issues they need to discuss with their mentees**
2. **Clarify their roles as teachers and role models in educating mentees about ethics**
3. **Manage the power dynamic inherent in the mentoring relationship**

Overview of Activities for the "Cultivating Ethical Behavior" Session

	Learning Objectives	Core Activities
1	Articulate the ethical issues they need to discuss with their mentees	Share ethics guidelines and discuss mentor roles in teaching ethics (Activity #1)
2	Clarify their roles as teachers and role models in educating mentees about ethics	Read and discuss three case studies: "Tweaking the Data," "Plagiarism?", and "A Big, Strong Guy" (Activity #2)
3	Manage the power dynamic inherent in the mentoring relationship	Read and discuss the case study "A Drive in the Country" (Activity #3)

FACILITATION GUIDE

Recommended Session for Cultivating Ethical Behavior (60 minutes)

Materials Needed for the Session:

- Table tents and markers
- Chalkboard, whiteboard, or flip chart
- Handouts:
    - Copies of introduction and learning objectives for Cultivating Ethical Behavior (page 105)
    - Copies of case studies: "Tweaking the Data," "Plagiarism?", "A Big, Strong Guy," and "A Drive in the Country" (pages 108–109)

Introductions (5 min)

TELL ▶ Review the introduction and learning objectives for the session.

Objective 1: Articulate the ethical issues they need to discuss with their mentees (20 min)

ACTIVITY #1: Discussion of Professional Societies' Ethics Guidelines (20 min)

ASK ▶ Ask mentors to share the ethics statement published by their disciplinary professional society.

DISCUSS ▶ Here are some possible guiding questions:

- How do these guidelines apply to your work?
- Which issues do mentors need to share with their mentees?
- How can mentors teach these ethics to mentees?
- What are good resources for teaching ethics?
- What responsibility do mentors have for making sure mentees keep a good notebook? Is this an issue of ethics?

Objective 2: Clarify their roles as teachers and role models in educating mentees about ethics (20 min)

ACTIVITY #2: Case Studies: Tweaking the Data, Plagiarism?, and A Big, Strong Guy

READ ▶ Distribute copies of the case studies for participants to read, and have them select which case to discuss first and second (and third, if time permits).

DISCUSS ▶ (10 min per study) Engage the entire group in discussion. You may want to record the ideas generated in this discussion on the whiteboard or flip chart. Use the guiding questions following the case study to facilitate discussion. Some additional guiding questions include the following:

- How can you teach a mentee good ethical behavior?
- The entire scientific enterprise depends on trust (and the ability to repeat experiments), so what is the ethical responsibility of a mentee to keep a detailed notebook?
- What is the responsibility of a mentor with respect to teaching a mentee how to keep a good lab notebook?
- How do new researchers learn ethical practices in their discipline? What is general and what is specific to the discipline?
- Are there special issues for learning ethical practices in multidisciplinary research? What are they?
- What ethical issues can arise due to the power dynamic between mentor and mentee?
- How can a mentor's reaction to unexpected news motivate or influence a mentee to make good or bad ethical choices? What is the issue or point of conflict?

Objective 3: Manage the power dynamic inherent in the mentoring relationship (15 min)

ACTIVITY #3: Case Study: A Drive in the Country

READ ▶ Distribute copies of the case study for participants to read.

ASK ▶ Ask mentors how power plays a role in their relationships with their mentees.

TELL ▶ Let participants know that being a mentor is accompanied, to some degree, by responsibility for another person's career and well-being. Because mentors are usually respected and admired by mentees, over whom mentors often have power, there is a power differential. Even if they feel that they have no power because they do not assign a grade or write letters of recommendation, the simple fact that their mentee is learning about research from them means that they are in the more powerful position in the relationship. As a result, some degree of dependence and need for approval often influences mentees' freedom about expression of ideas and choices.

DISCUSS ▶ Discuss the case with the entire group. You may want to record the ideas generated in this discussion on the whiteboard or flip chart. Use the guiding questions following the case study to facilitate discussion.

Postsession Assignments

TELL ▶ Ask participants to revise their draft mentoring philosophy to incorporate the issues discussed in this session. Have them bring two copies of their draft to the next session for peer review.

Case Study: Tweaking the Data

John is mentoring an undergraduate in his lab and has assigned her to collect data for one of the experiments in his dissertation. When the dataset is complete, he sits down to analyze it and finds his predictions completely disconfirmed. Dismayed, he calls her into his office and asks her to describe, in great detail, what she did when collecting the data. He wants to make sure that these anomalous results can't be more easily explained by mistakes in the lab. Their conversation lasts quite a while, but at the end he is still frustrated and puzzled by the data, and he sends her home so he can think about it some more.

Later, John is eating lunch in the cafeteria when he overhears his mentee talking to a friend of hers. Sounding very upset, she tells her friend, "I think John is mad at me," and describes their recent meeting. John is surprised to realize that his mentee took his questioning very personally. When John's mentee finishes venting, her friend replies, "If he's so mad, you probably did make a mistake somewhere. After all, he's the expert. Maybe you should tweak the data a little next time to keep him happy."

Guiding Questions for Discussion:

1. Who are the stakeholders in this case (individuals, institutions, public)?

2. What are the facts? What assumptions are you making about the situation?

3. What courses of action are possible? Which ones are preferable and why?

4. What, if anything, could have been done to prevent the situation?

Case Study: Plagiarism?

A team of four undergraduate researchers has been asked to research and write a literature review about a process they will be using during work in your lab. They have been given clear guidelines about your expectations for citations in the paper. When the review is handed in, you notice that some sections of the paper have language that seems too technical for the students to have written it, and a quick Internet search shows that these sections are copied verbatim from a published paper available online. You invite the team of students to your office to discuss the issue. What do you say?

Guiding Questions for Discussion:

1. Who are the stakeholders in this case (individuals, institutions, public)?

2. What are the facts? What assumptions are you making about the situation?

3. What courses of action are possible? Which ones are preferable and why?

4. What, if anything, could have been done to prevent the situation?

Case Study: A Big, Strong Guy

You are the graduate student mentor to a team of students working in the field on a research project that occasionally requires some serious manual labor. The PI for the project visits the research site often. When he is there and needs some physical work done, he always asks for "a big, strong guy" to volunteer to help him. A female student on your team volunteers after one of these requests, and your PI says, "Are you a big, strong guy? No!" He then turns away, still looking for someone to assist him. What do you do?

Guiding Questions for Discussion:

1. Who are the stakeholders in this case (individuals, institutions, public)?

2. What are the facts? What assumptions are you making about the situation?

3. What courses of action are possible? Which ones are preferable and why?

4. What, if anything, could have been done to prevent the situation?

Case Study: A Drive in the Country

A female undergraduate student and her male mentor were planning to attend the national meeting of the Society for Advancement of Chicanos and Native Americans in Science (SACNAS) in a city an 8-hour drive from their university. A few weeks before the meeting, the mentor went into the lab and suggested to the student that they drive to the meeting together. He said they could stay over in a hotel to break up the driving into two days, and it would still cost less than flying. He commented on how it was a very scenic drive, they might be able to collect some research samples along the way, and it would give them unbroken time to talk about research and her plans for graduate school. As the student hesitated, she saw all of her lab mates stealing curious glances at her while the mentor waited for an answer.

Guiding Questions for Discussion:

1. What is the power dynamic in the mentoring relationship and what factors create it?

2. How might power issues have affected the undergraduate student's choices? For example, if another undergraduate student had offered to travel together, would her response have differed? Why?

3. Even assuming that the mentor and student have a strong, trusting relationship, why might the student be uncomfortable in this situation?

4. How might the mentor have handled the situation differently?

5. Is it acceptable for mentors to travel with mentees of the opposite sex (or of the same sex in the case of a mentor who is known by the community to be gay or lesbian)? To professional meetings? For fieldwork? If not, does this, on average, disadvantage women students? Why?

Articulating Your Mentoring Philosophy and Plan

Introduction

Reflecting upon your mentoring relationships is a vital part of becoming a more effective mentor. This is especially important immediately following a mentor-training session so that you can consider how to implement changes in your mentoring practice based on the training. Reflection on your mentoring practice at regular intervals is strongly encouraged.

Learning Objectives

Mentors will have the knowledge and skills to

1. **Reflect on the mentor-training experience**
2. **Reflect on intended behavioral or philosophical changes**
3. **Articulate an approach for working with mentees in the future**

Overview of Activities for the "Articulating Your Mentoring Philosophy and Plan" Session

	Learning Objectives	Core Activities
1	Reflect on the mentor-training experience	Mentors engage in an open discussion of the knowledge and skills they have learned from the mentor-training sessions (Activity #1)
2	Reflect on intended behavioral or philosophical changes	Mentors reflect on components of mentoring and write about their mentoring practices before and after the mentor-training sessions (Activity #2)
3	Articulate an approach for working with mentees in the future	Mentors discuss philosophies and plans for working with a new mentee (Activity #3)

FACILITATION GUIDE

Recommended Session for Articulating Your Mentoring Philosophy and Plan (60 minutes)

Materials Needed for the Session:

- Table tents and markers
- Chalkboard, whiteboard, or flip chart
- Handouts:
  - Copies of introduction and learning objectives for Articulating Your Mentoring Philosophy and Plan (page 111)
  - Copies of the Mentoring Reflection Worksheet (page 115)

Objective 1: Reflect on the mentor-training experience (15 min)

ACTIVITY #1: Group Discussion of Lessons Learned from Mentor Training

ASK ▶ Have each person share with the group one or two ideas that stand out for them from the mentor-training sessions. These can include lessons learned, ideas that did or did not resonate with them, and so on. Once everyone has had a chance to share, open the discussion for additional comments. You may want to record ideas generated in this discussion on the whiteboard or flip chart.

Objective 2: Reflect on intended behavioral or philosophical changes (10 min)

ACTIVITY #2: Individual Written Reflection across Mentoring Themes

ASK ▶ Have each participant individually complete the Mentoring Reflection Worksheet. If there is not enough time to complete the writing activity, they may finish later.

NOTE ▶ Encourage mentors to edit their compact (if applicable) with these ideas. Another similar tool can be found in "*Nature*'s Guide for Mentors" by Adrian Lee, Carina Dennis, and Philip Campbell (*Nature* 447, 791–797, 2007).

Objective 3: Articulate an approach for working with mentees in the future (25 min)

ACTIVITY #3: Discussion of Ways to Begin a New Mentoring Relationship (10 min)

TELL ▶ Ask participants to imagine that they will begin mentoring a new undergraduate researcher next month.

DISCUSS ▶ Talk about the approach to this new situation with the entire group. You may want to record the ideas generated in this discussion on the whiteboard or flip chart. Guide the discussion using the following questions:

1. Specifically, what steps would you take to prepare for meeting with the new mentee in three weeks?
2. What will you do within the first month of the mentee's arrival?
3. What do you think is the most important thing you can do to start this new mentoring relationship off on the right foot?

ACTIVITY #4: Sharing of Mentoring Philosophies (15 min)

ASK ▶ Instruct participants to swap mentoring philosophies with each other and provide comments verbally.

DISCUSS ▶ Have participants discuss their reactions to the mentoring philosophies with the whole group. Some possible guiding questions include the following:

- What struck you about the mentoring philosophy you read?
- Where on the scale of visionary to detailed should a philosophy be?
- Where does your philosophy fall on the scale?
- How will you use your mentoring philosophy?

NOTE ▶ As part of the discussion or to wrap it up, be sure to stress the following points:

- A mentoring philosophy is a work in progress, and mentors should think about revising it as they go along.
- The mentoring philosophy could be included as part of a teaching philosophy.
- Mentors might consider using the philosophy as the basis for a handout to new mentees to establish and communicate expectations.

FACILITATION GUIDE

Final Thoughts (10 min)

Application of Research Mentor Training

TELL ▶ Be sure to share with participants that they are now qualified to provide research mentor training and can offer it at their future institutions if they so choose. They can obtain curriculum material at www.researchmentortraining.org and http://mentoringresources.ictr.wisc.edu. Let them know they will be receiving a follow-up evaluation. Evaluation instruments can be found at http://mentoringresources.ictr.wisc.edu.

TELL ▶ Encourage participants to use this training to their advantage in their professional careers. They should refer to this training, both their participation in it and their ability to teach it to others, in future job applications and grant proposals.

Mentoring Reflection Worksheet

For each mentoring theme, please list one or two approaches you have taken in the past and plan to take in the future.

Theme	Approaches you have used in the past	Approaches you intend to try in the future
Aligning Expectations		
Promoting Professional Development		
Maintaining Effective Communication		
Addressing Equity and Inclusion		
Assessing Understanding		
Fostering Independence		
Cultivating Ethical Behavior		

PARTICIPANT MATERIALS

Introduction to Facilitation

The following materials are designed to assist you in your role as facilitator of the research mentor-training curriculum. Specifically, these materials will help you guide the mentors during training sessions as they work through their thoughts and ideas and engage in self-reflection and shared discovery.

Roles of Facilitators

Importantly, the role of a facilitator is not to teach others how to mentor, but rather to guide them. As a facilitator, your role is to

- **Make it safe:** Take time to tell the group members that the research mentor-training sessions are a safe place to be honest about their ideas and feelings. Everyone's ideas are worth hearing. For it to feel safe, the content of the sessions must be treated respectfully and kept in confidence.

- **Keep it constructive and positive:** Remind members of your group to keep things positive and constructive. Ask the group how they want to deal with negativity and pointless venting. Remind them the training is about working together to learn, not complaining about the current situation or discounting the ideas of others in the interest of a personal agenda.

- **Make the discussion functional:** At the start of each session, explain the goals of the session to the group. Try to keep the group on task without rushing them. If the conversation begins to move beyond the main topic, bring the discussion back to the main theme of the session.

- **Give group members functional roles and responsibilities:** Assign or ask for volunteers to take notes, keep track of time, and report to the entire group at the end of the session. Functional roles help keep participants engaged.

- **Give all participants a voice:** In a group, there are likely to be issues of intimidation and power dynamics that can play out in ways that allow certain members of the group to dominate while others remain silent. At the start of the conversation, point out that a diversity of perspectives is an essential part of the process. Remind group members to respect all levels of experience. It's important that everyone's voice is heard.

General Notes on Facilitating a Group*

Each group takes on its own feel and personality based on the people in the group, your approach as a facilitator, and a host of external factors beyond your control. It helps if you are able to release your expectations for how a meeting or group should go, and instead focus on core aspects of the process. Your role as facilitator is to be intentional and explicit, while remaining flexible and not overly prescriptive. To a large extent it is up to the participants to take ownership of their own learning, especially since this training is designed for adults. Individual ownership, self-reflection, and shared discovery and learning will promote the deepest learning for this type of program.

As challenges and normal group dynamics surface, the group will look to you to fix problems. Part of your role is to help others see that they are also responsible for fixing problems. You can help them realize this by adhering to the following core ideas of group dynamics (and periodically reminding the team of them):

- Respectful interactions (listening, nonjudging, nondominating, genuine questioning, etc.) are essential.
- Relevant tangents that tie back to a central topic, issue, or question are fine, but don't let them derail the central purpose of the discussion.
- You need to keep moving ahead, but there is no need to push the schedule if the group needs time to reflect or slow down. If you slow down or skip something, you can anticipate participants will feel they are behind or missing out, so reassure them this is normal and the initial schedule is only a guide; there will be time to revisit topics if needed.
- If you try something and it doesn't go well, don't abandon it right away. Step back and think about what went wrong, talk to the group, learn from it, and try it again. It often takes a time or two to get the group warmed up to something new.
- Discomfort and silence are OK, but with a clearly stated context and purpose. Silence may seem like a waste of time in meetings, but it gives people a chance to think, digest, and reflect. Allow for a few silent breaks before, during, and at the end of each meeting.
- Make it easy, rewarding, and fun for people to participate, and encourage others to do the same for each other. Simple things like friendly reminders of meetings; providing coffee, tea, or snacks; and follow-up calls to check in with someone if they miss a meeting all send the message that you care and want to make it easy for individuals to participate.

---

* Adapted from *Creating a Collaborative Learning Environment Guidebook*, Center for the Integration of Research, Teaching, and Learning: http://www.cirtl.net/files/Guidebook_CreatingACollaborativeLearningEnvironment.pdf (accessed April 3, 2014).

Group Dynamics: Suggestions for Handling Challenges*

What do I do when no one talks?

- Have participants write an idea or answer to a question on a piece of paper and toss it in the middle of the table. Each participant then draws a piece of paper from the center of the table (excluding their own) and reads it out loud. All ideas are read out loud before any open discussion begins.
- Have participants discuss a topic in pairs for 3 to 5 minutes before opening the discussion to the larger group.

What do I do when one person is dominating the conversation?

- Use a talking stone to guide the discussion. Participants may only talk when holding the stone. Each person in the group is given a chance to speak before anyone else can have a second turn with the stone. Participants may pass if they choose not to talk. Importantly, each person holding the stone should share their own ideas and resist responding to someone else's ideas. Generally, once everyone has a chance to speak, the group can move into open discussion without the stone.
- Use the Constructive/Destructive Group Behaviors exercise in Chapter 1. Each participant chooses their most constructive and most destructive group behavior from a list (see page 15) and writes the two behaviors on the back of their name tag or table tent. Then participants share their choices with the larger group and explain why they chose those behaviors. This exercise also helps provide the group with a vocabulary so they may name these behaviors as they later note them in themselves and others. It provides a lighthearted and nonthreatening way that they can help each other stay on track.

What do I do when the group members direct all their questions and comments to me instead of their fellow group members?

- Each time a group member talks to you, move your eye contact to someone else in the group to help the speaker direct their attention elsewhere.
- Ask the participants for help in resolving one of your mentoring challenges. For example, ask them for advice on how to deal with an apathetic mentee. This helps the group members stop looking to you for the right answers and redirects the problem-solving and discussion focus to the entire group.

What do I do when a certain person never talks?

- Have a different participant initiate each discussion to ensure that different people have the chance to speak.
- Assign participants different roles in a scenario or case study and ask them to adopt a certain perspective for the discussion. For example, some participants could consider the perspective of the mentee, while others consider the perspective of the mentor.
- Try smaller group discussions (two to three participants per group), since some participants may feel more comfortable talking in smaller groups or without certain group members present.

---

* Adapted from Branchaw, J., Pfund, C., and Rediske, R. (2010), *Entering Research: A Facilitator's Manual: Workshops for Students Beginning Research in Science.* W.H. Freeman & Company.

What do I do when the group gets off topic?

- Have everyone write about the ideas they want to share on a given topic for 3 minutes. This short writing time will help participants collect their ideas and decide what thoughts they would most like to share with the group so they can focus on that point.
- Ask someone to take notes and recap the discussion at the halfway and end points of the session to keep the conversation focused.

About the Authors

Christine Pfund, PhD, is a researcher with the Wisconsin Center for Education Research at the University of Wisconsin–Madison (UW). Dr. Pfund earned her PhD in Cellular and Molecular Biology, followed by postdoctoral research in Plant Pathology, both at University of Wisconsin–Madison. For almost a decade, Dr. Pfund served as the Associate Director of the Delta Program in Research, Teaching, and Learning and the codirector of the Wisconsin Program for Scientific Teaching, helping to train future faculty to become better, more effective teachers. Dr. Pfund is now conducting research with several programs across the UW campus, including the Institute for Clinical and Translational Research and the Center for Women's Health Research. Her work focuses on developing, implementing, documenting, and studying research mentor-training interventions across science, technology, engineering, mathematics, and medicine (STEMM). Dr. Pfund coauthored the original *Entering Mentoring* curriculum and coauthored several papers documenting the effectiveness of this approach. Currently, Dr. Pfund is coleading two studies focused on the impact of training on both mentors and mentees and understanding specific factors in mentoring relationships that account for positive student outcomes.

Janet Branchaw is the Director of the Institute for Biology Education at the University of Wisconsin–Madison. She earned her BS in Zoology from Iowa State University and her PhD in Physiology from the University of Wisconsin–Madison. After completing postdoctoral research training and a lectureship in undergraduate and medical physiology at the University of Wisconsin–Madison's School of Medicine and Public Health, Dr. Branchaw joined the Institute for Biology Education. Her scholarship and program development expertise are in the areas of research mentee and mentor professional development and in the development and evaluation of interventions designed to support the success of first-generation, underrepresented minority and socioeconomically disadvantaged students. She is the lead author on the *Entering Research* curriculum and has led two National Science Foundation–funded undergraduate research programs to prepare diverse populations of students for graduate education: a 10-week summer Research Experiences for Undergraduates program that hosts students from around the country, and a three-year Undergraduate Research and Mentoring program. She leads the University of Wisconsin–Madison's Howard Hughes Medical Institute–funded "Foundations for Success in Undergraduate Biology" program. As the Director of the cross-campus Institute for Biology Education, Dr. Branchaw oversees development of innovative educational programs in educator professional development, K–12, undergraduate and graduate education, and science outreach and community engagement.

Jo Handelsman is a Howard Hughes Medical Institute Professor in the Department of Molecular, Cellular, and Developmental Biology at Yale University. She served on the faculty at the University of Wisconsin–Madison from 1985 until moving to Yale in 2010. Her research focuses on the genetic and functional diversity of microorganisms in soil and insect gut communities. She is one of the pioneers of functional metagenomics, an approach to accessing the genetic potential of unculturable bacteria in environmental samples for discovery of novel microbial products, and she recently served as President of the American Society for Microbiology. In addition to her research program, Dr. Handelsman is also known internationally for her efforts to improve science education and increase the participation of women and minorities in science at the university level. Her leadership in education led to her appointment as the first President of the Rosalind Franklin Society; her service on the National Academies' panel that wrote the 2006 report, "Beyond Bias and Barriers: Fulfilling the Potential of Women in Academic Science and Engineering"; her selection by President Barack Obama to receive the Presidential Award for Excellence in Science, Mathematics, and Engineering Mentoring; her position as cochair of a working group that produced the 2012 report to the President, "Engage to Excel: Producing One Million Additional College Graduates with Degrees in Science, Technology, Engineering, and Mathematics," about improving STEM education in postsecondary education; and *Nature* listing her as one of the "ten people who mattered" in 2012 for her research on gender bias in science.

The Entering Mentoring Book Series

The Entering Mentoring Book Series is a collection of curricula aimed at improving research mentoring relationships. The series includes training curricula for mentors and mentees at the undergraduate, graduate student, postdoctoral, and faculty levels across science, technology, engineering, mathematics, and medical disciplines. Authored by well-known science educators, the Series provides readings, resources, and active learning approaches that encourage mentors and mentees to reflect upon and improve their relationship. Each training curriculum includes an easy-to-follow, detailed facilitation guide for educators to use and adapt for their own settings.

Ongoing information and resources regarding mentoring are located at www.researchmentor training.org and https://mentoringresources.ictr.wisc.edu

Co-editors for the Entering Mentoring Book Series:
Christine Pfund, Wisconsin Center for Education Research, University of Wisconsin—Madison
Jo Handelsman, Department of Molecular, Cellular, and Developmental Biology and Center for Scientific Teaching, Yale University

For further information about the series, please contact:

Beth Cole
Acquisitions Editor, Life Sciences
W.H. Freeman & Company | Macmillan Higher Education
41 Madison Ave, New York, NY 10010
beth.cole@macmillan.com